GREEN THUMB GUIDES

cannabis
Everything You Need to Grow Marijuana Indoors and Outdoors

Kevin Oliver

ALPHA

ALPHA

Publisher: Mike Sanders
Editorial Assistant: Kylie McNutt
Art & Design Director: William Thomas
Cover Designer: Lindsay Dobbs
Book Designer: Rebecca Batchelor
Photographer: Allie Beckett
Illustrator: Laura Robbins
Compositor: Ayanna Lacey
Proofreader: Mira S. Park
Indexer: Johnna VanHoose Dinse

First American Edition, 2024
Published in the United States by DK Publishing
1745 Broadway, 20th Floor, New York, NY 10019

Copyright © 2024 by Kevin Oliver
24 25 26 27 10 9 8 7 6 5 4 3 2 1
001-334218-AUG2024

Library of Congress Catalog Number: 2023931547
ISBN: 978-07-440-7628-8

DK books are available at special discounts when purchased in bulk for sales promotions, premiums, fund-raising, or educational use. For details, contact: SpecialSales@dk.com

Printed and bound in China

Reprinted from *Idiot's Guides®: Growing Marijuana*

www.dk.com

MIX
Paper | Supporting responsible forestry
FSC™ C018179

This book was made with Forest Stewardship Council™ certified paper—one small step in DK's commitment to a sustainable future.
For more information go to www.dk.com/our-green-pledge

Note: This book contains information on cannabis, a controlled substance in much of North America and across the world. The possession, use, and cultivation of cannabis can carry heavy penalties. The publisher does not condone the use of any illegal substance. Readers should educate themselves about the laws in their jurisdictions before employing this scheduled substance. The author and publisher specifically disclaim any responsibility for any liability, loss, or risk, personal or otherwise, which is incurred as a consequence, directly or indirectly, of the use and application of any of the contents of this book.

This publication contains the opinions and ideas of its author. It is intended to provide helpful and informative material on the subject matter covered. It is sold with the understanding that the author and publisher are not engaged in rendering professional services in the book. If the reader requires personal assistance or advice, a competent professional should be consulted.

Contents

Part 2:
Introduction to Growing Cannabis 29

Part 3:
Indoor
Growing.............99

Part 4:
Outdoor Growing...........145

Part 5:
Extending Growing Seasons with Structures.........179

Part 6:
Breeding 191

Introduction

Cannabis has been cultivated for use by humans for thousands of years. However, it's only recently that cannabis growing has been legalized in certain states and countries. This has led many people to become interested in how to grow their own cannabis crop for personal medical and/or recreational use.

In this book, I'll teach you how to grow and maintain a small cannabis garden for your own noncommercial use. I'll cover the full spectrum of options available for growing cannabis for personal use—from indoor, climate-controlled systems; to open-air outdoor growing; to grow structures—as well as the entire growth cycle of the marijuana plant. You'll also receive both time-tested and modern methods for strain selection, disease and pest prevention, and proper plant nutrition. If you follow these tips and guidelines, you'll be well on your way to growing healthy, potent cannabis plants for personal use by responsible adults.

Disclaimer

Growing cannabis legally for personal use has been a long-sought-after but only recently earned right—and only in some jurisdictions. I advise you to educate yourself about the laws regarding personal, adult-use marijuana growing—and the number of plants allowed—in your state or country. Links to laws for U.S. states can be found at ncsl.org/research/health/state-medical-marijuana-laws.aspx and at NORML.org.

Acknowledgments

Thanks to the following organizations and individuals, without whose advice, guidance, cooperation, and efforts this book would not have been written: the National Organization for the Reform of Marijuana Laws (NORML) for its ongoing mission to legalize marijuana for use by responsible adults; Alison Holcomb of the American Civil Liberties Union (ACLU) for helping to lead the way in a new approach to marijuana policy; the designers, editors, and publisher at DK; Allie Beckett for her talented work as a photographer; and everyone else who provided technical expertise and advice during the writing of this book. Finally, many thanks to the individuals who have grown cannabis over the millennia for both their perseverance during times of prohibition and the resultant preservation of the cannabis gene pool, as well as for their continuing contribution to cannabis horticulture. To one and all, I raise a joint—smoke 'em if you've got 'em.

Special Thanks to the Technical Reviewer

Green Thumb Guides: Cannabis was reviewed by an expert who double-checked the accuracy of what's presented here. Special thanks are extended to Daniel Vinkovetsky.

About the Author

Kevin Oliver cofounded and operated Washington's Finest Cannabis (FineWeed.us), a legal cannabis farm licensed by the Washington State Liquor Control Board in 2014. He has been interested in gardening since childhood, when he began helping his grandmother in her rose garden, and has been growing cannabis for more than 20 years. He has been interviewed frequently as an industry expert by a variety of media outlets, including NBC's *Today*, NPR, Bloomberg, MSNBC, Forbes, and Al Jazeera America. Kevin is well known in the marijuana reform movement as a community organizer, law reform advocate, leader, cannabis farmer, and public speaker. He sits on the board of directors of the National Organization for the Reform of Marijuana Laws (NORML) and is executive director of the Washington State Affiliate of NORML. *Dope* magazine awarded his farm the 2015 Best Outdoor Grow in Washington State and awarded NORML the Best Cannabis Association in 2016 and 2017.

Part 1
Cannabis 101

Everything you need to know about cannabis is here, including information on its origins, species, strains, usage, and anatomy.

Cannabis: An Overview

Marijuana is the common name of a drug made from the cannabis plant. It has been grown and used around the world for thousands of years for everything from hemp products, to spiritual practices, to medicinal treatment, to recreation.

Geographic Origins

Cannabis originated in Central Asia along a series of foothills that extend from the base of the Himalayas to the base of the Hindu Kush mountain ranges. These nutrient-rich areas allowed the plant to flourish. It was then brought to Korea by Chinese coastal farmers around 2000 BCE. Over time, invaders and explorers brought the plant to the Middle East, Russia, Germany, Britain, Africa, South America, and finally North America. It reached the United States via Mexico in the twentieth century, during the Mexican Revolution. Cannabis can now be found in every civilized country and inhabitable corner on Earth.

History of Cannabis Usage

The first use of cannabis for its effect on humans can be traced to around 2727 BCE. in the writings of Chinese emperor Shen Nung, who is credited with documenting the healing and meditative qualities of cannabis. Its popularity has spread across the millennia, gaining mainstream cultural popularity in the United States during the 1920s and 1930s. As of the writing of this book, marijuana possession, production, processing, adult-use retail sales, and/or personal cultivation has been legalized in 22 U.S. states and territories, and 37 U.S. states have legalized some type of medical marijuana use. In addition, several countries allow possession and consumption. Please review the laws in your state or country before attempting personal cultivation.

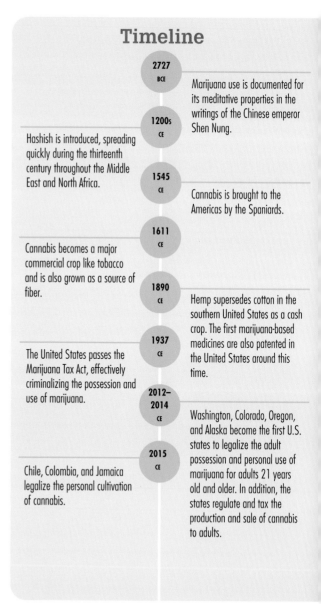

Timeline

2727 BCE
Marijuana use is documented for its meditative properties in the writings of the Chinese emperor Shen Nung.

1200s CE
Hashish is introduced, spreading quickly during the thirteenth century throughout the Middle East and North Africa.

1545 CE
Cannabis is brought to the Americas by the Spaniards.

1611 CE
Cannabis becomes a major commercial crop like tobacco and is also grown as a source of fiber.

1890 CE
Hemp supersedes cotton in the southern United States as a cash crop. The first marijuana-based medicines are also patented in the United States around this time.

1937 CE
The United States passes the Marijuana Tax Act, effectively criminalizing the possession and use of marijuana.

2012–2014 CE
Washington, Colorado, Oregon, and Alaska become the first U.S. states to legalize the adult possession and personal use of marijuana for adults 21 years old and older. In addition, the states regulate and tax the production and sale of cannabis to adults.

2015 CE
Chile, Colombia, and Jamaica legalize the personal cultivation of cannabis.

Marijuana Trivia

In 2008, the world's oldest cannabis stash was found inside a 2,700-year-old tomb in the Gobi Desert. Tests proved the marijuana possessed potent psychoactive properties, increasing speculation that it was used recreationally even then.

Anatomy of a Cannabis Plant

Before you start growing cannabis, it's important to familiarize yourself with both its basic and male and female anatomy. This can give you a working knowledge of the functions of each part, as well as help you determine the sex of the plant.

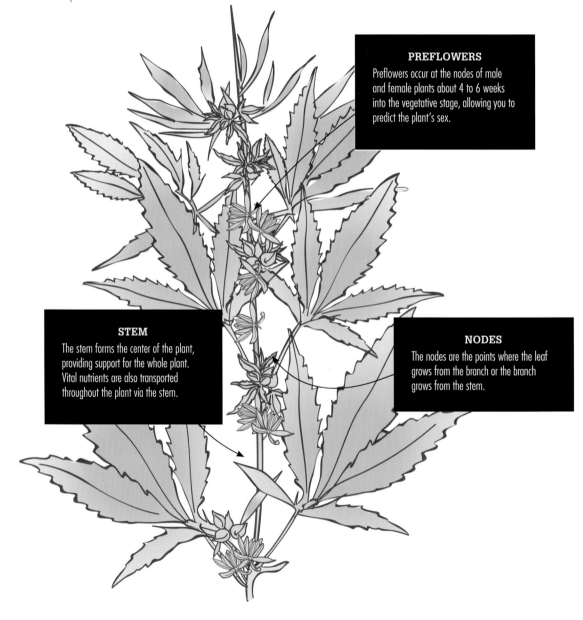

PREFLOWERS
Preflowers occur at the nodes of male and female plants about 4 to 6 weeks into the vegetative stage, allowing you to predict the plant's sex.

STEM
The stem forms the center of the plant, providing support for the whole plant. Vital nutrients are also transported throughout the plant via the stem.

NODES
The nodes are the points where the leaf grows from the branch or the branch grows from the stem.

LEAVES

The large leaves sprouting from the nodes along the branches provide the primary source of energy for the plant through the process of photosynthesis. Smaller leaves near and around the bud are generally considered usable marijuana trim (excess snippings of leaves from the buds used for extractions, tinctures, hash, and edibles). This is due to them being covered in sticky, resinous trichomes, which contain the main sought-after ingredient, tetrahydrocannabinol (THC).

BUDS

The usable flower of female cannabis is often referred to as the *bud*, or *calyx*. Buds are formed from up to thousands of individual flowers that grow from the tip of the stems, the tips of the branches, and at the nodes (where leaves meet branches and where branches meet the stem). Clusters of buds are called a *cola*; a main cola forms at the very top of the plant, while smaller colas form along the budding sites below.

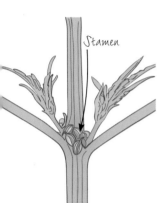

Stamen

MALE PARTS

The pollen-producing reproductive organs of a male flower are called *stamen*. Male marijuana stamen appear as a cluster of three to five or more small, green- or yellow-seed-sized preflowers.

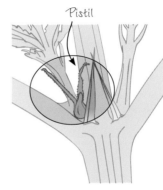

Pistil

FEMALE PARTS

The reproductive organs of a female flower consist of white, hairlike structures called *pistils*. These serve to collect pollen from the male plants. However, pistils typically don't have much of an effect on a bud's potency and taste.

Species of Cannabis:
Indica and Sativa

There are three known species of cannabis: indica, sativa, and ruderalis. However, the two most common species are indica and sativa.

Cannabis Indica **and** *Sativa*

Indica (*Cannabis sativa forma indica*) delivers more of a body high than other species of cannabis, causing more of a relaxed, sleepy feeling. Sativa (*Cannabis sativa L*) offers more of a "head" high that allows people to feel more energetic and creative. Indica tends to have a higher tetrahydrocannabinol-to-cannabidiol (THC-to-CBD) ratio, while sativa tends to have a higher CBD-to-THC ratio. All species of cannabis are believed to have originated in Central and South Asia, but indica has adapted to survive in hotter, dryer climates, such as the desert and some mountainous regions. Sativa grows in lusher, more humid environments, such as jungles and rain forests.

CANNABIS INDICA

Differences in Appearance
Visible differences between indica and sativa are seen in the leaves and overall size and shape of the plant. Indica leaves have shorter, broader, and fewer fingers than sativa leaves, which are usually longer and skinnier. Indica grows more shrublike, with mature colas resembling large pinecones. Sativas grow much taller, almost like a vine, with long and cylindrical mature colas resembling a forearm or the end of a baseball bat. Both species can be grown indoors, but indicas usually fare better than sativas—something to consider when deciding which type to grow, indoors or out.

Regulating Humidity Levels for Sativas

Sativas in bloom seem to thrive in slightly warmer and more humid environments. The ideal humidity level should be between 40 and 50 percent and never above 65 percent once the flowers have developed. Otherwise, the risk of the plants developing mold, powdery mildew, or other debilitating issues significantly increases.

CANNABIS SATIVA

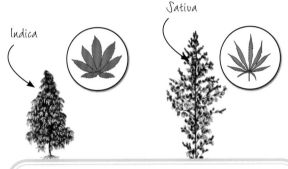

Indica

Sativa

Differences in the Vegetative Stage

In the vegetative stage, there isn't much difference between the species. However, keep in mind that most sativas will continue to grow rapidly in flower.

Differences in the Flowering Stage

The flowering times—the amount of days required for a plant to complete the flowering process—are different from strain to strain. While indicas can finish flowering in as little as 45 days and as much as 65 days, most strains take about 8 weeks to finish. Sativas take a little longer, usually finishing in about 9 to 11 weeks; however, some sativa strains can take even longer. In fact, some nonhybridized landrace sativa strains can take up to 20 weeks to flower!

Species of Cannabis:
Ruderalis and Hybrids

Due to the federal illegality of cannabis, no one is truly sure how many species of cannabis there are. However, sativa, indica, and ruderalis are typically considered the three main species of cannabis, with top strains of each species sometimes combined to create hybrids.

Cannabis Ruderalis

Although it's not nearly as popular as indica and sativa, ruderalis has become more prevalent in hybrid plant species in recent years due to its auto-flowering characteristics. Originating from Central Russia, it has a similar chemical profile to hemp and contains very low levels of THC, until it is crossed with high-THC indica or sativa plants.

CLASSIFICATION

The name ruderalis comes from the word *ruderal,* a term used by botanists to describe hardy, nondomesticated plants. So any breed of cannabis that has the ability to adapt to extreme environments and avoid human cultivation has been classified as ruderalis. Only recently have growers begun to utilize the characteristics of the ruderalis plant to influence new hybrid varieties.

PLANT CHARACTERISTICS

This stalky species of cannabis only grows to a height of 1 to 2½ feet tall (30.5cm to .75m) and produces small, chunky buds. But what really differentiates the ruderalis from indica and sativa is its flowering cycle. Unlike the other cannabis species, ruderalis's flowering cycle is induced according to its maturity and not by the photoperiod (seasonal changes in night length), a process known as *auto-flowering.* This auto-flowering can begin in as little as 21 days.

Bumper Crops

Through the process of genetic manipulation, ruderalis has helped implement bumper crops. Like the known gardening technique, bumper crops yield an unusually productive harvest. As this relates to cannabis, they are comprised of auto-flowering plants and, due to their smaller size, work great as supplemental plants in a garden that has a little more room to grow.

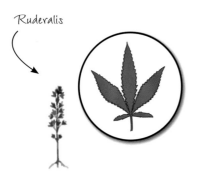

Ruderalis

Hybrids

Today, almost every marijuana strain has been hybridized as growers have combined their top indica, sativa, or ruderalis strains to create new strains with the best aspects of both parents. It's estimated that at least 90 percent of the world's marijuana being consumed is a hybrid strain. One well-known hybrid is afghanica, or afgooey. Once erroneously identified as indica, this strain is indica-predominant with a potent and sedating high. Great for indoor or outdoor growing, it can grow between 5 and 6 feet (1 and 2m) tall, depending on the setting. Another well-known type of hybrid is an auto-flowering strain. As you might expect, this comes from cross-breeding a strain with ruderalis. Very popular with growers, this hybrid is easy to grow due to the short time from seed to harvest and its ability to flower when mature rather than when seasonal changes happen.

CANNABIS RUDERALIS

AFGHANICA

Strains

There are more cannabis strains today than ever, and with more and more strains being crossbred, spliced, and grafted daily, the potential combinations of strains are virtually endless. The differences in strains can be seen visually, as well as in the smell, taste, and overall effect. You can also find differences in the growth patterns and length of the growth cycle from strain to strain.

The ability to refine a particular strain to carry more or less of particular attributes has resulted in a vast assortment of genetics being available on the cannabis market. Once a strain has been developed, it's usually named by the breeder and marketed by describing the particular characteristics of that strain. With the demand continuing to grow for new strains, the crossbreeding of genetics has become a very interesting dynamic in the cannabis industry. Breeders who are able to develop new genetics will one day be able to patent and claim those strains as their own. Here, I have compiled a list of varying strains from various breeders.

9 POUND HAMMER
Composition: 80% indica, 20% sativa **Crosses:** Gooberry × Hells OG × Jack the Ripper **Taste and smell:** Sweet grape and lime flavor and aroma **Property:** Sedating

AFGHAN OG
Composition: Indica-dominant hybrid **Crosses:** Afghan Kush × OG Kush **Taste and smell:** Earthy smell and taste **Property:** Sedating

AFGHANI
Composition: 95% indica **Crosses:** Unknown **Taste and smell:** Sweet taste with a pungent, earthy aroma **Property:** Sedating

BERRY WHITE
Composition: 100% indica **Crosses:** Blueberry × White Widow **Taste and smell:** Blueberry taste and smell **Property:** Energetic

BLUE HAWAIIAN
Composition: 30% indica, 70% sativa **Crosses:** Blueberry × Hawaiian Sativa **Taste and smell:** Berry taste and aroma **Property:** Energetic

BLUE LAVENDER
Composition: 60% indica, 40% sativa **Crosses:** Blueberry × Lavender **Taste and smell:** Lavender and berry taste and smell **Property:** Sedating

BLUE MOON
Composition: 50+% sativa **Crosses:** Afghani Kush × Blue Dream **Taste and smell:** Clean taste and smell with hints of blueberry **Property:** Energetic

BLUEBERRY
Composition: Indica-dominant hybrid **Crosses:** Afghani × Thai × Purple Thai **Taste and smell:** Sweet berry taste and a pungent blueberry aroma **Property:** Sedating

CACTUS
Composition: Indica-dominant hybrid **Crosses:** Afghani × Northern Lights **Taste and smell:** Earthy taste with a bit of a citrus smell **Property:** Energetic

CBD CRITICAL
Composition: Indica-dominant hybrid **Crosses:** Critical Kush × Ruderalis **Taste and smell:** Strong earthy aroma and taste **Property:** Sedating

CHERNOBYL

Composition: 80% sativa, 20% indica **Crosses:** Trainwreck × Jack the Ripper × Trinity **Taste and smell:** Lime flavor with a lime sherbet smell **Property:** Energetic

DAY TRIPPER

Composition: Indica-dominant hybrid **Crosses:** Sensi Star × Medicine Man × Master Kush **Taste and smell:** Fresh, earthy taste with a floral, sweet aroma **Property:** Energetic

DEVIL'S LETTUCE

Composition: 70% indica **Crosses:** Shiskaberry × Great White Shark **Taste and smell:** Sweet and spicy taste and smell **Property:** Energetic

DURBAN POISON

Composition: 100% sativa **Crosses:** African Sativa **Taste and smell:** Earthy, sweet, and piney smell and flavor **Property:** Energetic

DUTCH TREAT WHITE

Composition: 80% indica, 20% sativa **Crosses:** Unknown lineage **Taste and smell:** Sweet, fruity smell and taste with pine and eucalyptus leaves **Property:** Sedating

GOLDEN STAR

Composition: Sativa-dominant hybrid **Crosses:** Sensi Star × Malawi Gold **Taste and smell:** Amazing menthol and pine taste with floral and spice aromas **Property:** Energetic

GORILLA GLUE 4

Composition: Sativa-dominant hybrid **Crosses:** Chem's Sister × Chocolate Diesel **Taste and smell:** Sour diesel and chemical glue smell with a coffee taste **Property:** Energetic

JAGER

Composition: 80% indica, 20% sativa **Crosses:** Hindu Kush lineage **Taste and smell:** Black licorice smell and taste **Property:** Sedating

KANDY KUSH
Composition: Indica-dominant hybrid **Crosses:** OG Kush × Trainwreck
Taste and smell: Sweet, earthy, and citrusy smell and taste **Property:** Sedating

MAUI WOWIE
Composition: 70% sativa, 30% indica **Crosses:** Classic Hawaiian sativa
Taste and smell: Sweet pineapple taste and smell **Property:** Energetic

NORTHERN LIGHTS
Composition: 95% indica, 5% sativa **Crosses:** Afghani × Thai Landrace
Taste and smell: Earthy pine taste and a pungent, sweet, and spicy aroma
Property: Sedating

OG KUSH
Composition: Indica-dominant hybrid **Crosses:** Chemdawg × Hindu Kush
Taste and smell: Pine and sour lemon taste and scent **Property:** Energetic

PERMAFROST

Composition: Sativa-dominant hybrid **Crosses:** White Widow × Trainwreck
Taste and smell: Fresh pine aroma and taste **Property:** Sedating

PREDATOR PINK

Composition: 60% indica, 40% sativa **Crosses:** Plushberry × Starfighter
Taste and smell: Earthy, sweet berry aroma and taste **Property:** Energetic

PURPLE CHEMDAWG

Composition: 75% indica, 25% sativa **Crosses:** Chemdawg × Granddaddy Purple
Taste and smell: Earthy, sweet taste and smell **Property:** Sedating

PURPLE HINDU KUSH

Composition: 100% indica **Crosses:** Hindu Kush × Purple Afghani **Taste and smell:** Grape taste with a subtle, earthy aroma **Property:** Sedating

SNOOP'S DREAM

Composition: 50% indica, 50% sativa **Crosses:** Blue Dream × Master Kush
Taste and smell: Sweet blueberry aroma and flavor with pine aftertaste
Property: Sedating

SOUR DIESEL

Composition: Indica-dominant hybrid **Crosses:** Chemdawg 91 × (Mass Super Skunk × Northern Lights) **Taste and smell:** Pungent, diesel-like aroma and taste
Property: Energetic

THIN MINT GIRL SCOUT COOKIES

Composition: 50% indica, 50% sativa **Crosses:** Durban Poison × OG Kush
Taste and smell: Sweet, minty taste and aroma **Property:** Sedating

Part 2
Introduction to Growing Cannabis

Cannabis can grow just about anywhere with minimal resources, but it can also be cultivated, either indoors or outdoors, under controlled conditions to yield a much higher-quality and higher-quantity crop. In this part, I go over the basic stages of growth and what's important about each. I then finish with an important question: do you want to grow indoors, outdoors, or in a grow structure?

Why Grow Marijuana?

There are many reasons why growing cannabis plants in a home garden may be a viable option for you. The following takes you through four common reasons why growing your own rather than relying on outside sources is ideal.

1 You Can Grow What You Want
The smells, tastes, and effects of cannabis are influenced by terpenes, the primary components of essential oils that are present in most plants and flowers. A combination of terpenes makes up a terpene profile. More than 200 terpenes are found in marijuana plants, but only a handful are regularly prevalent in quantities large enough to affect the potency and smoking quality of marijuana. When combined, the terpenes and cannabinoids create an "entourage effect," which leads to an overall greater effect than any of individual available compounds. The most prevalent terpenes found in cannabis are as follows:

> **Limonene,** also found in citrus

> **Myrcene,** also found in hops, lemongrass, mango, and thyme

> **Pinene,** also found in pine

> **Linalool,** also found in lavender

> **Caryophyllene,** also found in pepper and cotton

Individual preferences for certain strains of marijuana vary from person to person. However, modern terpene testing of marijuana can determine the exact profiles of the plants you prefer. Based on that, you can grow cannabis that provides you with a consistent source of the strain you like best.

2 You Have Total Control Over the Product
Much like growing tomatoes or brewing your own beer, growing cannabis at home allows you total control over your own garden. Cultivating plants for personal use ensures you know exactly what's being done to and used on your plants, from roots to flowers. Because home gardens aren't subject to the standard methods and practices of commercial production, you have the option of growing marijuana that's free of both chemicals and pesticides.

3 You Save Money
The costs associated with growing your own marijuana could save you hundreds or even thousands of dollars over the course of a year. While there's an initial investment to start growing, this is dependent upon your needs and budget and can be quite minimal. For instance, consuming $1/2$ gram a day at $20 a gram would cost $3,600 over the course of a year, while a simple home grow could easily be set up for only a few hundred to a thousand dollars.

4 You Can Enjoy the Growing Experience
Growing your own cannabis is an environmentally sustainable, personally rewarding, psychologically satisfying, and recreationally enjoyable experience. Gardening, in and of itself, is a fantastic way to spend your time and attention—especially when choosing to nurture a plant with the sole purpose of enjoying its ultimate reward.

Disclaimer

The legality of marijuana varies state by state. It is your responsibility to know the laws in your state and the maximum number of plants you can grow for personal use. My assumption in this book is that your cannabis home grow will be legal; will be for responsible adult personal use only; and will include no more than the maximum number of plants allowed by law in your state, providence, or country.

The Life Cycle of Pot

Cannabis is an annual plant, meaning it can go through its entire life cycle within a year. The life cycle includes propagation or germination, vegetation, flowering, and harvest.

Propagation and Germination

The beginning of marijuana's life cycle can begin two ways: from a seed or from a cutting of an existing plant. If starting from a seed, you first need to germinate it, which involves placing it in a damp napkin in a warm, dark place and then waiting for the seed to pop open and produce a tail. If you're using a cutting, it needs to propagate or clone, which means you have to entice the cut to produce roots. You can propagate a cutting in several ways, such as placing a clone in presoaked rock wool cubes or plugs using a propagation device that sprays water on your cut at timed intervals.

Vegetative Stage

The vegetative stage is the main growing phase in which the seed or cutting becomes the plant. This 4- to 6-week stage is where cuts can be taken and the plants can be "beefed up" prior to going into the flowering or bloom stage. The vegetative stage can be prolonged almost indefinitely indoors by keeping the plants under light for 18 to 24 hours per day. When cultivating outdoors, many growers will keep their plants indoors in pots during the vegetative stage so as not to trigger flowering until the plants have reached a desired size. Once plants have reached the desired size, remembering that they will continue to grow in the flowering stage, you can put your plants into bloom.

Flowering Stage

Flowering can take anywhere from 6 to 10 weeks, depending on the strain and your growing technique. The flowering stage is where you really need to focus on the daily needs of your plants. Pests will become more of a threat, as will any type of mold or mildew. Also, pay attention to the timing of the nutrient regimen throughout flowering, as it will vary. During the last week or two, it's always wise to water your crops with some type of flushing agent or even plain, pH-balanced water to flush out any remaining nutrients. The trichome development during this stage determines when you should harvest. This development takes a strain-specific amount of time, and harvesting prematurely will result in a lesser-quality flower.

Harvest

At the end of the plant's life cycle is the harvest. This is when the plant has completed the flowering stage and is in its optimal state to take down and begin the drying and curing process. Curing—the slow process of drying and decarboxylation that converts cannabinoids to a usable, psychoactive form—is probably the most misunderstood step of the entire cultivation process. The smell, stickiness, and overall potency of the plant are heavily affected by properly curing the dried flowers.

What Cannabis **Needs to Live**

Cannabis needs basically what most plants need to live: healthy roots, a strong stem, and green leaves. As a general rule of thumb, marijuana thrives under the same basic garden conditions as tomato plants. As a grower, you control the variables that influence how well the plant grows.

Soil

A good soil delivers nutrients and drains water properly while providing a good foundation for the roots and plant stalk. (I talk more about soil in the next section.)

Water

Water is essential for the life of a plant. Some municipal drinking water systems may have levels of sodium, chlorine, chloramine, and fluoride that aren't conducive to plant growth. If this is the case for you, you can purchase distilled water or employ a reverse osmosis system.

Nutrients

Among many supplemental nutrients, nitrogen and phosphorus are essential for cannabis growth. Nitrogen provides healthy vegetative plant growth, while phosphorus provides the nutrients necessary for healthy, robust flowers in nice, tight buds.

Light

Providing as much light as possible to your marijuana plants—without burning them from excessive heat—ensures they'll utilize nutrients and water to their maximum potential. In general, your plants must have at least 12 hours of light a day, depending on what stage of growth they're in.

Temperature

An ideal temperature for growing marijuana indoors is between 70°F and 80°F (21°C and 27°C), give or take a few degrees. Outdoors, temperatures can sometimes vary widely. Precautions—such as a heater or a shade cloth—need to be taken to prevent extended exposure of marijuana plants to temperatures at or below 32°F (0°C) or above 90°F (32°C).

Humidity

The ideal humidity for marijuana plants is between 40 and 70 percent, depending on the growth stage of your plants. Seedlings and plants in the vegetative stage can thrive in climates with higher humidity; however, as flowers appear, the ideal humidity is around 40 percent to prevent mold.

Airflow and Circulation

Proper airflow will recirculate all of the air in an enclosed grow room or covered outdoor grow space in about 5 to 10 minutes. Fresh air should be brought in and stale air exhausted at an adequate rate, while proper temperature and humidity levels are maintained.

Ideal **Soil Composition**

You can use many different grow mediums for your marijuana plants, but the most common for indoor and outdoor growers is soil. Soil is the easiest grow medium for beginners to use and requires very little in the way of investment and research compared to other grow mediums. However, your soil should have certain characteristics to ensure your plants grow effectively.

Soil pH Level

One consideration for your soil is its pH. Cannabis plants grow best in soil with a pH between 5.5 and 7. Anything outside of that, and your plants won't get the essential nutrients they need, leading to stalled growth and even plant death. For outdoor growing, ensure the pH is in that range before planting. For indoor growing, soil mixes at your local gardening store will include pH values on them; check those numbers closely to find one that will keep your plants in the right range.

Drainage Quality

You also want to make sure the soil is loose and light enough to drain well while retaining moisture and air. When plants are in soil, they tend to hunt for water. With the right watering schedule, you can make this work to your advantage by making the roots search for food in well-draining soil, leading to stronger, healthier plants overall. When it comes to air retention, keeping the temperature of the soil from being too cold or too warm helps your roots thrive and produce the desired results. You want something that can keep your plants' root zone between 60°F and 70°F (15°C and 21°C).

Monitoring Indoor Soil

When growing indoors, remember to monitor the temperature, moisture level, and pH of your soil to ensure you're not causing the plant feeding system to become counterproductive. Overwatering is the easiest way to kill your plants, and overfertilizing will burn and possibly kill them as well.

Nutrients Needed in Soil

To supply your plants with nutrients, you'll want to add fertilizers that have the right ratio of NPK: nitrogen, phosphorus, and potassium. These generally come in two- or three-part mixtures that allow you to adjust each part at various stages of growth. Clearly, the three-part blends are easier to dial in the right ratios. You can then add supplements that will enhance certain characteristics of each individual strain in both the vegetative and flowering stages. Silica, humic acids, rooting hormones, carb-loaded supplements, and shooting powders are just some of the nutrients you can add to your soil that will bring out the best in your crops.

Making Your Own Soil

While you can find premixed bags of soil for growing, you can more easily ensure you get the right concentration of nutrients for your plants if you make your own. To make your own soil, you need two things: base soil and fertilizer. A base soil is simply a soil without additives. An ideal kind of base soil is organic soil with ingredients such as peat moss, worm castings, and perlite.

Once you have your base soil, you can supplement it with fertilizer. For instance, humic and fulvic acids are organic substances you can add to help your plants absorb the nutrients. The ratio of these nutrients will change as the plant goes through the various stages of its life cycle. Liquid and dry fertilizers can be bought premade at your local hydroponic supply or gardening store, or you can make them yourself. Most of your local hydroponic supply stores will have some type of formula to follow, and a few manufacturers even have nutrient recipes on their websites.

Rock wool

Rock wool, coco coir, and peat moss are just some of the mediums in which you can grow your marijuana plants.

Coco coir

Peat moss

How to Make Your Own
Grow Medium

Growing your own marijuana allows you to control the type of medium or soil you choose to grow it in, with the ingredients affecting the flavor and aroma. Generally, the more organic your medium, the better tasting your buds will be.

Kevin's 2:1:1 Ratio

This mix provides a good foundation for your plants and an efficient balance of nutrients while providing proper drainage and moisture retention.

2 parts Pro-Mix HP (a peat-based soil, with perlite and beneficial mycorrhizae fungi)

1 part FoxFarm Ocean Forest (contains a plethora of fine marine ingredients in addition to worm castings, bat guano, and forest humus)

1 part medium to large perlite aggregate (optional, to increase drainage capabilities)

1 Measure out Pro-Mix HP, FoxFarm Ocean Forest, and perlite aggregate (if using) using a 5-gallon (19 liter) bucket and place into a 50-gallon (189 liter) container.

2 Mix soil by hand, with a shovel, or using a soil mixer machine until it's well blended.

These are general guidelines for making your own soil. Pay attention to the manufacturer's recommendations for what works best for your specific soil.

From Scratch

Base soils that lack nutrients can also be supplemented to create a great medium for growing marijuana. A good, organic soil will provide all of the nutrients your cannabis plant needs to grow from the seedling stage through the harvest stage, while maintaining the proper pH to ensure nutrient uptake. If you'd like, you can later add dry supplements to the soil during plant growth and liquid supplements to the water when you water the soil. You can also foliar feed your plants by spraying water supplemented with nutrients onto the leaves.

1 **START WITH A BASE SOIL**
In the middle of a large tarp or a large 50-gallon (189 liter) container, mix your base soil.

2 **ADD NUTRIENT SUPPLEMENTS**
Add any of the nutrient supplements listed previously per the manufacturer's instructions.

3 **COMBINE THE BASE SOIL AND SUPPLEMENTS**
Mix your base soil and supplements thoroughly by hand, with a shovel, or using a soil mixer machine.

You can then monitor the pH of your soil medium by adding enough water to make it muddy and then testing it with a pH meter. The pH for your soil should be 6.2 to 6.5. Use a pH up or pH down solution to adjust the pH to the required range.

Ideal Fertilizers for Organic Soils

Alfalfa meal, bat guano, coco fiber, compost, Epsom salts, fish emulsion, fish meal, peat moss, pumice, soybean meal, and worm castings are some of the best fertilizers to add to your base soil.

Hydroponics: No Soil Required

Growing without soil, or growing hydroponically, allows a faster growth cycle through a more controlled method of nutrient delivery to the plants.

The basic concept is that the roots are exposed to an exact nutrient regiment and, in most applications, have the opportunity to dry prior to the next watering cycle, leading them to grow faster than in soil. I'll discuss the several types of hydroponic systems, as well as their pros and cons, later in the book. Right now, let's take a look at why you should or shouldn't use hydroponics to grow your marijuana.

Why Grow Hydroponically

Faster growth cycles equate to more frequent harvests, which is why most indoor growers utilize the methods of hydroponics. You can also manipulate the nutrient uptake, forcing plants to develop particular, desired characteristics. For instance, when watering in soil, your crop will get watered, at most, every day—but more than likely every other day or so. However, if you're growing in a hydroponic system, your crop can be watered several times per day, allowing you to have a more exact nutrient regimen for your plants.

Another benefit of hydroponics is the ability to mix in a flushing agent more frequently. This helps remove salt-leaching crystals that build up on the roots over time, ensuring your plants take up the necessary nutrients they need. Plus, in a hydroponic system, you're far less likely to have pests, meaning you don't have to use pesticides. This lack of pesticide use eliminates something that can potentially kill or destroy your plants' growth.

Why Not to Grow Hydroponically

A major concern in the hydroponic grow room is mechanical failure. Everything from lights, to fans, to air conditioners, to water chillers and pumps can fail. Should any of these mechanisms go, plants can face dangerous and potentially destructive circumstances without the protection of soil. However, many experienced growers implement backup devices, such as temperature sensors, so preventative measures will be in place to compensate.

Another disadvantage of growing via hydroponics is cross-contamination due to the plants sharing the same water source. This is especially prevalent in recycled watering systems—should one plant become contaminated, they all risk contamination. And when plants in hydroponics take a substantial hit, they will rarely come back 100 percent. So keep these risks in mind when deciding whether to grow in soil or in a hydroponic system. Basically, soil is much more forgiving, but hydroponics is much faster. Choose wisely!

The Importance of **pH**

One of the most important things to keep in mind when feeding your plants is the pH level. In particular, many plant nutrient deficiencies are related to an acidic pH, which can cause deficiencies in many of the essential elements needed to grow cannabis, such as calcium, magnesium, nitrate-nitrogen, and phosphorus. Therefore, knowing what pH level is necessary and how to test it is essential.

Alkalinity vs. Acidity

The pH scale ranges from 0 to 14, with 7 considered neutral. Anything above 7 is considered alkaline, while anything below 7 is considered acidic. The ideal pH for plants in a hydroponic system is between 5.8 and 6.2; for plants in soil, it's between 5.5 and 7. Typically, your plants' pH will rise in a hydroponic system as nutrients are absorbed, which are acidic in nature. Meanwhile, your plants' pH in soil remains relatively stable due to the dirt buffering the roots of your plants from fluctuations.

As the pH changes between alkaline and acidic, different nutrients will be absorbed or locked out. This could potentially lead to slowed growth or even plant death. The pH can be adjusted by adding an acid (such as sulfur) or a base (such as ground limestone, wood ashes, or oyster shell) to your water, or even a premixed pH up or down solution. These can be found at your local garden supply or hydroponics store.

How to Test the pH Level

There are many devices on the market to test the pH level of your water. However, seeing as this is a very important part of the growing process, it's wise to invest in a better-than-average pH tester. Measuring and adjusting the level will vary between soil and hydroponic applications. After mixing your nutrients for your grow medium, the pH must be adjusted to not only ensure efficient uptake of the nutrients, but also to make sure your plants' growth isn't hindered by an improper pH level.

Testing the pH in hydroponic applications is much easier than in soil, as the pH level shouldn't change much in water once it's adjusted. In soil, the pH level of the runoff water can be very different than the pH level prior to watering. This will vary according to what you have in your soil already, as many premade soils will contain fertilizers. That means you'll want to test your pH prior to watering, as well as test the runoff water after you water to figure out what your root zone pH level is. For instance, if your nutrient water starts at 6 and your runoff drop is at 5, your medium is roughly 4. If you raise the pH of your water, it will raise the pH of your medium.

To test the pH level of the water in a hydroponic system, first dip a pH meter in calibrating solution and adjust the meter accordingly so it's properly calibrated. (This ensures the most accurate reading.) Next, insert the pH meter into the water. Based on the results you get, you can add a pH down or pH up solution to the water to adjust the pH.

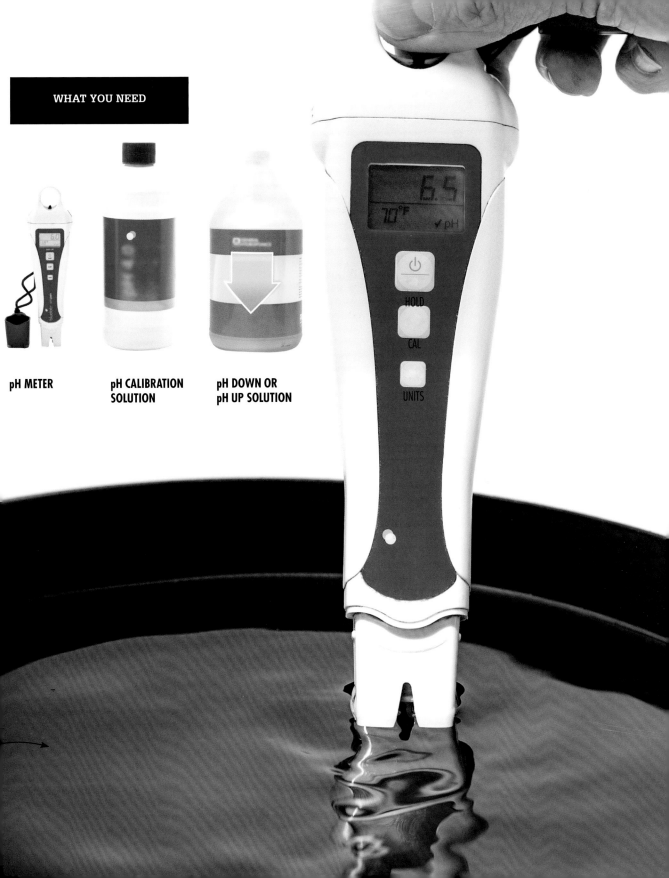

pH METER

pH CALIBRATION SOLUTION

pH DOWN OR pH UP SOLUTION

Water

Water is essential to most life as we know it. This is why monitoring your water is of the utmost importance to the success of your garden. The pH, total dissolved solids (TDS), temperature, and nutrients added in relation to your water will determine the quality and quantity of the harvest at the end of your plants' life cycle.

pH

As you learned previously, pH is adjusted by using an acid to lower the level or a base to raise the pH level. Different water pH levels are required for different growth mediums and strains. For instance, rock wool cubes used for cloning should be soaked in a water solution adjusted to 5.5. In hydroponic applications, your pH level should start at 5.8 and be adjusted once it reaches 6.2. In soil, while your ideal pH level will vary depending on the strain you're growing, it will be between 5.5 and 6.5. Your meters are good at letting you know when the pH in your water needs to be adjusted, so it's important to learn how to read what they're telling you. The wrong pH can cause serious damage to your plants and inhibit their growth.

TDS and Parts per Million

Total dissolved solids (TDS) are the total amount of minerals, salts, or metals dissolved in a given volume of water. It's basically the measure of anything that's not H_2O in your water. TDS can include anything from sewage and fertilizer to rock and lead. Knowing the TDS of your water can help you troubleshoot any problems with your plants related to nutrients.

TDS is expressed in parts per million (ppm), which refers to the amount of particulates that are in the water per million parts. In terms of measuring TDS, you can find out the concentration of a nutrient you've added to your water (for instance, nitrogen) and cross-reference it with the amount of the nutrient required by your plants to see whether they're nutrient deficient or overabundant. TDS can also indicate chlorine or any other unwanted particulates you need to remove from the water.

Reverse Osmosis

When it comes to getting rid of unwanted particulates, some gardeners prefer to strip the water completely through a process of filtration called *reverse osmosis* (*RO*). The downside to filtering water through RO is the expense and waste, as even the best RO systems still put out at least a 1:1 ratio of clean water to wastewater.

Temperature

The temperature of your water is very important to control, especially when growing in hydroponic or aeroponic applications. In soil, while the temperature of the water isn't as detrimental, keeping the temperature in the desired range is always ideal. Your water should always stay in the range of 65°F to 80°F (18°C to 27°C). At temperatures above 80°F (27°C), nonbeneficial bacteria begin to appear and can cause many problems for your crop, including entire crop loss. At temperatures below 65°F (18°C), your plants can go into shock. A water chiller can help with the more-likely high temperatures you'll face.

Nutrients

As I discussed with soil composition, nutrients are your fertilizer in water. You either build your own soil by adding fertilizers or add fertilizers to your water during feeding. This enhances certain characteristics of your plants and entices them to grow faster. Nitrogen, phosphorus, and potassium (NPK) are the three main nutrients needed for photosynthesis and therefore optimal growth. These will vary during each stage of growth, and additional nutrients can be supplemented as well to further enhance your plants' quality and yield amount.

OVERWATERED

UNDERWATERED

When watering your cannabis plants, they can sometimes get too much or too little of a good thing. Overwatering or underwatering your plants can cause a host of problems, such as a slowing of growth due to lack of oxygen for the roots and a greater potential for disease. So how do you know if you've overwatered or underwatered? Overwatered plants will have curling and yellowing leaves, while underwatered plants will wilt and have dry-looking leaves. To fix overwatered plants, allow the soil to dry and then poke holes in the soil to increase the oxygen. For overwatered plants in a hydroponic system, let the roots dry and then decrease the frequency of watering. To fix underwatered plants in soil (as underwatering in a hydroponic system is not an issue), simply water your plants more frequently with plain, unenriched water.

Acquiring Seeds

Seeds can be acquired from a variety of sources, such as online, through fellow growers, in magazine advertisements, in seed catalogs, and even from friends who found some in their marijuana stashes.

Purchasing and Exchanging

Purchasing seeds can be as easy as going online and searching for "cannabis seeds." But knowing which companies are reputable can take a bit more research. Seek out companies that offer seeds from award-winning breeders around the world. Many marijuana seed exchange websites also have a large selection of seeds from brands that have been vetted for reputation, quality, and customer satisfaction.

Feminized Seeds

While you can acquire a mix of seeds, feminized seeds are ideal for getting the most out of your grow. Feminized seeds are created by altering the hormones and stressing a female cannabis plant to create pollen without any male genetic material, thereby guaranteeing the females it pollinates will produce female seeds. This means you don't have to worry about killing off male plants before they fertilize female plants. Whether you're purchasing or producing your own, be sure the feminized plant and the females to be pollinated are mature plants of the same strain.

Feminized seeds are produced in three ways: silver thiosulfate, colloidal silver, and rodelization.

Silver thiosulfate. The most common method, applying this substance causes the female plant to turn into a male through hormonal stress over a period of 4 or 5 weeks. Pollination of females can then occur.

Colloidal silver. This is a complex procedure that accomplishes the same result as the silver thiosulfate method except with pure silver, distilled water, and a 9-volt battery. However, it's not recommended for small home grows, and plants coated with silver shouldn't be consumed or smoked.

Rodelization. This all-natural way involves allowing the plants to live in the flowering stage much longer than usual, thereby relying on nature to jump-start a natural self-pollination process. Though not as reliable as the previous methods, the completely natural nature of this method appeals to growers.

Because laws regarding seed acquisition vary from state to state, be sure you're up on current laws before you start.

Germinating Seeds

The process of seeds sprouting roots to grow a healthy plant is known as *germination*. The simplest methods for germination are soaking seeds overnight, starter cubes, direct planting, and the paper towel method.

Soaking Seeds

Soaking seeds in warm water for 1 to 1½ days (no longer) can facilitate the germination process. Once the seeds have opened and the roots are visible, move them to one of the other methods listed, being careful not to damage the roots.

Starter Cubes

Starter cubes have holes in them to accept seeds and generally provide a good environment for germination. Keep the cubes moist but not soaking wet during germination, which should take 3 to 5 days.

Direct Planting

Direct planting in your soil medium allows seeds to germinate as they would naturally. Plant them about ¾ inch (2cm) deep in moist but not wet soil and allow to germinate for 3 to 5 days.

Paper Towel Method

This is the easiest and simplest method for seed germination, in addition to providing a relatively maintenance-free process to sprout your highly valued seeds.

1 Wet a paper towel with warm (not hot) water. Squeeze out any excess moisture and place the towel on any plate. Set seeds on the damp paper towel and wrap them in it. Allow to sit at a comfortable room temperature for 2 or 3 days.

2 Check the towel two or three times a day (every 8 to 12 hours or so) to be sure it remains damp and to add warm water as necessary. You can help keep the moisture in during this time by covering the paper towel with another plate, a bowl, or some plastic wrap.

3 After the 2 or 3 days, seeds should start to show signs of opening to expose a small white tap root. Make sure not to damage the root after it has sprouted, prior to transplanting.

Planting **Seedlings**

If a seed has sprouted a root, it's referred to as a seedling. Once you have a seedling, you then need to plant it in whichever medium you have chosen for growing your plant, taking care not to damage the fragile root.

In a Soil Medium Indoors

A seedling can be planted in your chosen soil medium for an indoor garden.

1 Prepare about ½ to 1 pint (.25 to .5 liter) of soil.

2 Carefully place the seed in a small hole at a depth of about ¾ to 1 inch (2 to 2.5cm) in your soil medium.

3 Check the medium daily, making sure to keep it moist to foster robust root growth.

After the plant has sprouted into a seedling about 5 to 6 inches (12.5 to 15.25cm) tall, it will be ready for transplant to a larger container or to be moved outdoors. You can also judge this by looking to see if healthy roots are emerging from the bottom holes on the pots.

In Dirt Outdoors

1 Create a small hole about ¾ to 1 inch (2 to 2.5cm) deep in your prepared soil or dirt medium outdoors.

2 Moisten the soil. Carefully place the seed in the hole, covering it with the soil medium.

3 Check the soil daily, making sure to keep it moist.

In a few weeks, you'll have a nice seedling with several leaves and a healthy root system. Once the root systems have developed, the growth of your cannabis plant can be as much as a few inches (cm) a day.

In a Root Medium (Hydroponic)

Hydroponic systems will generally utilize a small cube of rock wool, kept moist, to hold seeds for their growth into seedlings. After sprouting into seedlings, the rook wool can be transferred to your hydroponic system, where it will absorb water from the surrounding medium and grow into a healthy plant. Some hydroponic systems utilize larger blocks of rock wool to receive your seedling starter cube, which is then fed a constant supply of water. However, seedlings grown in rock wool can be employed in a variety of hydroponic systems.

Rock wool is created by melting rock and sand and spinning it into wool-like fibers.

Vegetative Stage

Once your clones or seedlings have sprouted roots and begun to grow, they have started the transition into the vegetative stage. Plants take in nutrients from the leaves and the roots; now that the roots have started to show, your plants will grow much faster. The average vegetative stage lasts around 4 to 6 weeks.

Your Plants' Environment

As with all growing stages, the environment your plants are grown in during the vegetative stage will largely determine their success. Lighting selection, growing technique, and strain are all important considerations, as well as temperature and humidity.

Nutrients

During the vegetative stage, the intent is to make your plants as robust and strong as possible. That way, they'll be able to sustain the weight created by the buds once they begin to flower. Supplying the right nutrients at this time is a big part of making this happen successfully. Silica and humic acids are two nutrients you can add to your base grow medium to help strengthen the branches.

Techniques to Enhance Output

Making the most of the vegetative stage will give you the maximum output in the flowering stage. You can work the plant using different techniques during this stage to train your crop for optimal production.

TOPPING

This technique involves cutting off the primary cola of your plants, which allows your plants to share resources that would normally go toward the growth of their primary colas. After topping, the next tier of colas grow larger and absorb more of the environment's resources, leading to more buds come harvest time. And you don't have to waste those primary colas; you can then use them for cloning.

SUPER CROPPING

Otherwise known as *crimping,* this technique forces the plants to allow the lower branches to catch up growth-wise to the branches you have crimped. To perform this technique, gently crush the stem walls of the upper branches and bend them to a horizontal angle. This will slow the growth of that particular part of the branch, diverting resources to the other areas and leading to bushier and healthier plants. Super cropping can be done in conjunction with topping or on its own.

LOW-STRESS TRAINING

Low-stress training (LST) is very similar to super cropping, except the branches are tied with string or twist ties or weighted down with fishing weights instead of being gently forced into a horizontal position. This exposes the inner parts of your plants to more light, which leads to more developed growth.

LOLLIPOPPING

With lollipopping, you basically trim off the bottom third of your plants' branches so all their resources are diverted to the top of the plant. It can be done in the vegetative stage or in the very beginning of the flowering stage (though to do so too far into the latter will inhibit your plants' performance). Removing these branches can also improve air circulation around the lower reaches of your plants, where humidity can often build up, especially after watering. This makes your plants more robust and should result in a higher yield and quality. This technique is definitely not for all strains, so try it on a couple plants and compare the outcome at harvest.

Pruning at the Vegetative Stage

Pruning your marijuana plants throughout the vegetative stage can have many positive effects. Everything from growth patterns, to the weight of a plant's yield, to overall plant health are affected by pruning.

Just One Tool Needed

A clean, sharp pair of small pruning scissors works best for pruning. However, some gardeners may use their fingers to pinch a growing or emerging shoot off the plant. In every case, care should be taken to not harm any of the surrounding plant tissue.

Before   After

Getting the Desired Effect

Pruning can be done on small and large cannabis plants. However, it should not be done on very young plants. You should wait until you have at least a couple five-bladed leaves on a plant before you begin any pruning exercises.

Making the choice to prune your plants should not be a random act. Are you pruning to control the height or shape of your plants? Are you pruning to increase your harvest yields? Are you pruning to allow your plants better light and airflow? Once a pruning regime has begun, it should be maintained for maximum benefits.

PRUNING FOR SHAPE

Pruning the top of the plant controls the overall plant height. It also causes the plant to grow outward. If you're growing indoors, this may be necessary due to the height constraints of the room. Allowing a plant to grow out rather than up can facilitate light penetration through the canopy, especially when grown under lights.

PRUNING FOR YIELD

Maximizing the yield of your harvest is a good goal to consider. When pruning to maximize yield, the emerging shoots—which look like tiny leaves—are removed at the nodes where the fan leaves meet the branches, leaving about a half-dozen untouched near the tips of the branches. This directs the plant to use its energy to help develop the remaining ones into supporting healthy preflowers and large buds.

At the nodes where the fan leaves meet the branches, emerging shoots such as the ones shown here are pruned to yield larger buds at the tips of the branches.

THINNING THE PLANT

Light and airflow are important to a plant's health. Pruning for thinning allows more light and air to flow through a plant. Thinning can include removing small and weak branches, as well as dead or yellowing leaves.

To help improve airflow and light penetration, you can cut away some of the fan leaves from your plants. These images show how a dense plant is trimmed to expose leaves in different areas of it.

Fertilizing at the Vegetative Stage

Marijuana plants in the vegetative stage tend to thrive when provided with certain key nutrients. These nutrients are available in most gardening or hydroponic supply stores, premixed in dry or liquid form.

Nitrogen, Phosphorus, and Potassium

Nitrogen, phosphorus, and potassium (NPK) make up the basic macronutrients of most fertilizer mixes. Marijuana plants in the vegetative stage thrive when given fertilizers higher in nitrogen and lower in phosphorus. Phosphorus is essential for plant growth, but nitrogen is vital to photosynthesis during this time. The ratio of each nutrient will be listed on the label of the fertilizer; for example, a label reading of 5-1-1 has a 5 percent makeup of nitrogen and a 1 percent makeup of phosphorus and potassium, respectively.

Different strains of cannabis can have different demands for fertilizers, with some more sensitive to fertilizers than others. If you're unsure how much fertilizer to use, start with half of the recommended amount. Generally, if your plants look good (green leaves with no discoloration, as well as straight, firm leaves and stems), they're getting the nutrients they need.

Alfalfa meal is an all-natural fertilizer with a 3-1-2 makeup that helps replenish the soil to keep your plants healthy.

Bat guano, which has a 10-3-1 makeup, promotes rapid growth of your plants.

Liquid or Dry Fertilizers

Fertilizer mixes come in liquid or dry form. While liquid fertilizers tend to be slightly more expensive, some find them to be more convenient when mixing with water for use as either a foliar spray or a root system waterer.

Dry fertilizer mixes can sometimes be added directly to the soil medium at the base of the plant or, like liquid fertilizer, mixed with water to be used for either foliar feeding or watering of the roots. This allows them to release their nutrients over time with every watering, eliminating the need to add fertilizers to your water.

Whichever you decide to use, follow the manufacturer's instructions for mixing ratios and application, as too much fertilizer will harm a plant. Most require a fraction of a teaspoon of dry fertilizer or 1 ounce (30ml) of liquid fertilizer per 1 gallon (4 liters) of water.

Required Equipment

Standard equipment used for fertilizers include measuring spoons, liquid measuring cups, a 1- to 5-gallon (3.75- to 19-liter) bucket for mixing or watering, and a spray bottle for foliar feeding (2-gallon [7.5-liter] sizes work well).

Foliar Feeding

Foliar feeding is a good way to fertilize plants to ensure healthy growth. Turn off your high-intensity lights during spraying and while plants are wet to prevent burns from water drops, which tend to act as a magnifying glass in directing light.

Keeping **Vegetative Plants** Healthy

Maintaining a consistent environment is a main contributor to the health of your garden during the vegetative stage and in general. This means the temperature, relative humidity, and light cycle should always stay as strain specific as possible. If you're growing multiple strains with different genetic makeups, you'll want to create separate environments for each strain, provided this is financially and logistically feasible for you.

Aside from keeping your garden in an environment conducive to successful cannabis cultivation, you can introduce a few supplemental things to ensure your plants remain as healthy as possible throughout the vegetative stage.

Foliar Spraying

Foliar spraying during the vegetative stage allows your plants to absorb essential nutrients through their leaves and provides extra protection from predators. Most gardeners use four types of foliar sprays: fertilizers, pesticides, fungicides, and water.

Fertilizers. These are used as a supplemental form of feeding and generally promote node sight development, as well as the overall growth of the plant. Most fertilizers can be started as soon as your plants are in the vegetative state and can be used once or twice a week up to 3 weeks into flowering.

Pesticides. If used correctly, organic pesticides are all you'll ever need to keep unwanted visitors from entering your garden. Pesticides are used once a week.

Fungicides. Organic or synthetic fungicides help treat plants infected with fungus and prevent the spread of a fungal disease in your crop. Most gardeners implement it as a part of their weekly routine.

Water. If you don't have a humidifier, plain water can be sprayed anytime you feel the room needs more moisture.

BURNED TIPS

Make sure you're only doing your foliar sprays for the proper amount of time, or you may see burned leaf tips, which is a sign of nutrient burn.

External Devices That Promote Healthy Plants

If you feel more humidity is needed, you can introduce humidifiers during the vegetative stage. These can be set up with a controller that will enable you to maintain a perfect relative humidity (RH) percentage in your grow room. Fans are also a staple in any grow room during the vegetative stage. Moving air not only makes for a stronger plant, it allows for a more thorough mixture of the temperature and moisture levels in the room.

Transplanting

If the leaves on your plants begin to curl unexpectedly but they're large and healthy otherwise, the vegetative stage might be the time to transplant your plants into bigger containers. This fairly simple process helps rehabilitate your leaves. When transplanting, it's wise to have a substrate (a base in which the plant lives) on the bottom, such as expanded clay pellets (hydroton), to allow the ends of the roots to dry out. This will inhibit root rot and allow the roots to grow stronger. After transplanting, your plants should begin to show signs of recovery within a week. (You'll learn more about transplanting later in this part.)

Avoiding Windburn

Be sure your fans aren't set too high and aren't directly and constantly blowing on your plants. Doing so could lead to windburn, which can dry out and stress your plants.

CURLED, YELLOWING LEAVES

This indicates a nutrient deficiency, making foliar spraying essential for your plants.

DROOPING LEAVES WITH SPOTTING

When you see this, it's a sign of plant infestation. Fungicides and pesticides can help you avoid and treat this problem.

Transplanting

Transplanting can occur several times during the vegetative growth stage, depending on how much time you choose to allow the root system and plant to grow. This method encourages your plants to continue growing, as they would otherwise stall in their original containers.

When to Transplant

Your plants should be transported during the vegetative stage of growth. At this time, your plants will be going through rapid leaf growth, with more and more "true leaves"—the type of leaves most associated with the look of marijuana—appearing. The stem will also be firm enough to grasp gently without causing damage.

Avoiding Dry Spots

To ensure consistent soil texture and to avoid any dry spots, break up any clumps of soil when preparing your plants for transplant.

Transplant Method

The following takes you step by step through transplanting your cannabis plants.

1 Choose containers to be transplanted into that have holes in the bottom to allow proper water drainage. Prepare them by filling them with your desired soil medium, leaving room to set the incoming transplants. Water the soil medium to ensure moisture for the incoming root systems.

2 Gently remove the rooted plants to be transplanted from their existing containers. If possible, keep the entire root systems intact while holding the moist soil around them together. (The containers should act as a mold, as the root systems tend to hold the soil intact.)

Prolonging the **Vegetative Stage**

Plants can grow almost indefinitely in a vegetative state when kept in a controlled environment. But why would growers choose to prolong the vegetative stage of their plants? The following are some pros and cons of doing this.

Pros

The most obvious reason why you might intentionally prolong the vegetative stage is to increase the size of your plants. The larger the root system is, the larger the plant will be. Another reason you might prolong the vegetative stage is to keep some plants as "mothers," which can then be used as a continuous source from which to take clones (discussed more later in this part).

Cons

While yields can increase with larger plants, they will take longer to grow and will require more lighting equipment if grown indoors. Also, the longer plants are kept in the vegetative stage, the larger they get. As plants get larger, they require more space. This requires you to control your plants' growth so they don't grow by leaps and bounds.

As you transplant, take care not to disturb the root systems of your cannabis plants.

3 Set the transplants into the prepared containers. Fill the containers with more of the soil medium, surrounding the entire transplanted root systems. Ensure your marijuana plants are planted at a depth that allows the top of the transplanted soil to be equal to or slightly lower than the top of the soil in the new container. Also, make sure any branches aren't buried under or touching the soil.

4 After transplanting is complete, water the surrounding soil and root systems. Because settling of the soil can occur during watering, you can add more soil as necessary. When watering your transplant, use a spiral motion to cover the entire surface area of your soil medium, from the stem to the edge of the container. Continue watering in this fashion until water is running out of the holes in the bottom of your planting container.

Identifying the **Gender of a Plant**

Cannabis plants can be male or female. Female plants produce flowers high in THC, while male plants produce pollen sacs in small flowers with very little THC. Therefore, it's important to determine what the sex of your plant is.

What to Do with Male Plants

If you simply want to grow female plants, dispose of the male plants in your trash or compost pile to prevent them from pollinating the female flowers, leading to prized buds turning to seed. However, if pollinating for seed is your goal, you can keep the male plants. Just make sure the male plants are isolated from the female plants to prevent any unwanted pollination.

Why Sex Your Plants?

Knowing the sex of your plants allows you to maintain females for the potent THC in their flowers or to maintain males for pollen production.

At What Stage to Do It

Plant sex can be determined before flowering begins by examining preflowers at the nodes of vegetative plants after 3 or 4 weeks of vegetative growth. Plants begin to flower when they're deprived of light for at least 12 hours a day.

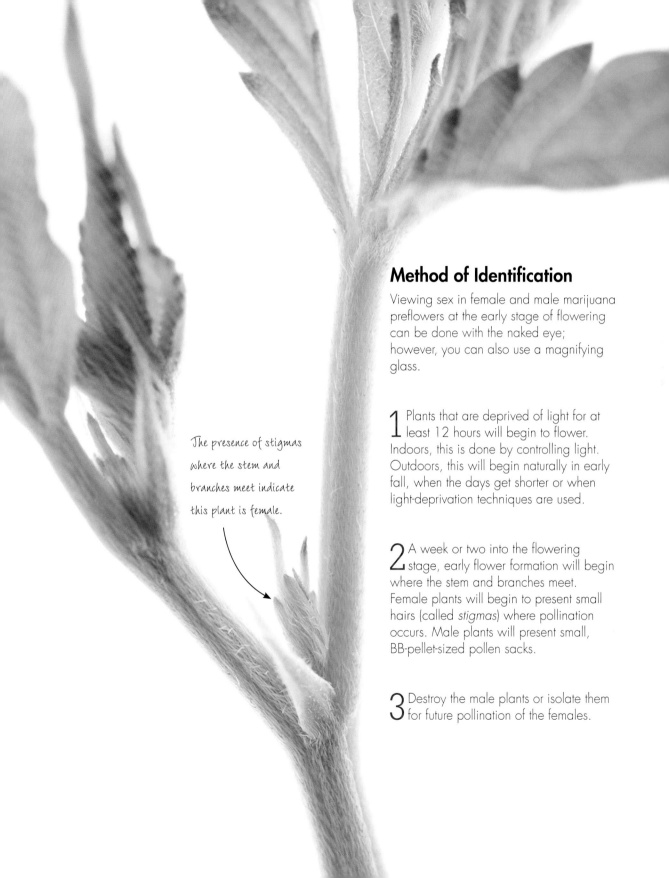

The presence of stigmas where the stem and branches meet indicate this plant is female.

Method of Identification

Viewing sex in female and male marijuana preflowers at the early stage of flowering can be done with the naked eye; however, you can also use a magnifying glass.

1 Plants that are deprived of light for at least 12 hours will begin to flower. Indoors, this is done by controlling light. Outdoors, this will begin naturally in early fall, when the days get shorter or when light-deprivation techniques are used.

2 A week or two into the flowering stage, early flower formation will begin where the stem and branches meet. Female plants will begin to present small hairs (called *stigmas*) where pollination occurs. Male plants will present small, BB-pellet-sized pollen sacks.

3 Destroy the male plants or isolate them for future pollination of the females.

Why Grow from **Clones?**

Not to be confused with breeding—in which you combine a male and female plant through pollination to create offspring with the characteristics of each parent—cloning involves taking part of a male or female plant and placing it in a rooting medium so you can start growing another plant with the exact same characteristics as the "mother" plant (called that even though the original you start with could be male).

Growing More of the Same

If you want to continue growing a particular type of cannabis, cloning what you already have saves you the expense of buying more of the same seeds. Another nice thing is that, in female plants, cloning doesn't diminish potency, meaning your cloned plants will have the exact same effect as the mother plant with no drop-off in what the original provides. In terms of male plants, cloning helps preserve your favored genetics for later crossbreeding.

No Worrying About Determining Gender

Because you're using only one plant in the cloning process, all of your plants are guaranteed to be the same gender—so plants from a male mother plant will be male, while plants from a female mother plant with be female.

Saving Time

Another nice advantage of cloning is you don't have to germinate the clones. They simply have to take root in the rooting medium before they enter the vegetation stage. This means the time from planting to harvesting is reduced, allowing you to harvest more cannabis crops than if you had to continually grow your plants from seed.

Clone Warning

You can clone a mother plant more than once, but take care not to clone the same one too many times. Doing so could stress or shock the plant, leading it to stop growing or die. To avoid this, simply cycle out mother plants.

How to **Clone**

To clone marijuana, you cut the plant's stem close to the branch and remove all but three or four of the top leaves from the cutting, leaving 3 to 6 inches (7.5 to 15.25cm) of the stem. You then moisten the bottom of the stem, dip it in a rooting hormone, and set it in a cloning tray.

AT WHAT STAGE SHOULD YOU CLONE?
Take clones from healthy vegetative plants that have not yet entered the flowering stage, as hormones present in flowering plants prevent clones from taking root as easily as in the vegetative stage.

OPTIMAL CONDITIONS FOR CLONING
Optimal conditions for cloning are air temperatures 5 to 10 degrees less than the rooting medium. Humidity can be decreased to 80 percent after the first week.
   Air temperature: 75°F to 80°F (24°C to 27°C)
   Humidity: 80 percent
   Lighting: Fluorescent (high-output T5 recommended) or low-wattage high-intensity discharge (HID; 400w). Lighting can be provided 24 hours a day.

HOW LONG IT TAKES
7 to 18 days

EQUIPMENT
Grow light assembly with T5 or MH bulbs
Heating mat (optional)
Sharp cutting utensil, such as a knife or scissors
Rooting medium cubes, such as peat moss or rock wool
1 cloning tray and humidity dome
16 oz. (.5 liter) cup or container
1 gallon (3.75 liters) clean water
Rooting hormone (recommended)

Cloning Tips

- Use the healthy middle branches of the plant for cuttings.

- When taking cuttings, the tips of the branches typically aren't used.

- Cut close to the branch, being careful not to cut the branch itself.

Cloning Method

1 Hang the lighting and place the heating mat (if using) on the floor or a table. Sterilize your tools and working area with an alcohol-, vinegar-, or vodka-and-water solution.

2 Place 1 rooting medium cube for each cutting in a cloning tray and moisten with water.

3 Choose your mother plants. Clone mother plants should be healthy plants at least 6 to 8 weeks old in a vegetative (nonflowering) state that are as clean of fertilizers, fungus, insecticides, molds, and pests as possible.

4 Cut each mother plant close to the branch.

5 Remove all but the top three or four leaves, leaving 3 to 6 inches (7.5 to 15.25cm) of the stem.

6 Cut off the leaf tips to lower transpiration and evaporation.

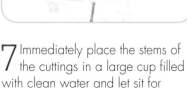

7 Immediately place the stems of the cuttings in a large cup filled with clean water and let sit for 1 hour to soften the tissue.

8 Cut the stem of the mother plant at about a 45-degree angle.

9 Split the stem and scrape the outer surface area about 1 inch (2.5cm) from the bottom. Scraping aids the stem in absorbing the rooting hormone.

12 Once all cuttings are inside, cover the cloning tray with the humidity dome. Keep the rooting medium between 75°F to 80°F (24°C to 27°C) and the air temperature of the cuttings a few degrees cooler. The humidity should be as close to 100 percent as possible for the first week and then 80 to 85 percent after that.

10 Dip the bottom of the stem in a rooting hormone (for instance, gels, liquids, or powders).

11 Place the rooting medium in a cloning tray. Moisten the medium and make a small hole in the center of it to accept the cutting. Place the cutting in the medium, being careful to preserve the rooting compound on the stem.

13 Mist the cuttings once or twice daily, adding water to the cloning trays as needed to keep the rooting medium moist. The root medium and cuttings should be moist but not soaked with water.

14 After 1 week or longer, small white roots should appear at the bottom of the stem. The cuttings may also grow in height and produce new leaves. Allow the roots to grow to at least 1 inch (2.5cm) to ensure they're healthy prior to transplant.

Warning

Slow-rooting cuttings and wilting or browning leaves on cuttings can indicate a moisture imbalance, a weak cutting, a disease, or pests. These should be removed from the cloning tray for quarantine or destruction.

Using the Cloning Process to **Sex a Plant Grown from Seed**

Once you have successfully sprouted seeds, you'll need to determine the sex of your plants, if their sex is unknown. To determine the sex of your plants via the cloning process, you must keep track of clones of the same plant genetics, which you can do using the following three-step process.

You can place your rooted clones in plastic cups or similar containers with their grow medium.

1 **LABEL AND TRACK CLONES**
Once your plants have grown for at least 4 to 6 weeks, cut and root at least two clones from each older plant you wish to sex. Label each clone to keep it clearly associated with the parent plant it came from. You'll then keep at least one labeled clone in the vegetative stage and at least one labeled clone in the flowering stage.

Before determining the sex of a plant, you must force it into a flowering stage. Use a simple box or a sack for light deprivation.

2 FLOWER A CLONE FOR SEX

After approximately 2 weeks of at least 12 hours of darkness during the same time period of each day, the flowering clone will begin to exhibit visible signs of gender. Either female stigma (small white hairs that receive pollen) or male pollen sacs (about the size of a bb pellet) will have begun to form at the nodes, where the leaf stems meet the branches.

Place a box or sack over a clone to deprive it of light for 12 hours a day.

3 GENDER SELECTION

Unless your goal is to pollinate female plants for seeds, once you've determined the sex of your flowering-stage clone, you should destroy the males. Use your labels to track its cloned brother or sister in the vegetative stage. After you've tracked down the ones labeled with the same genetics, remove and destroy any male clones in both the early stages of flowering, as well as those male clones with the same label still remaining in the vegetative stage. This will leave only female clones that can be cloned indefinitely, providing you with a guaranteed source of female plant genetics to be used in the flowering stage and produce the prized marijuana buds.

Flowering Stage

Flowering can begin once the plant has established roots. This typically happens in the fall outdoors, which can be mimicked with lighting indoors. Depending on the strain, the flowering stage can last on average from 6 to 10 weeks.

Nutrients in Bloom

Enhancing the quality of your crop is not only dependent on the ideal environment, but also your nutrient regimen. During the first few weeks of flowering, you'll want to continue on a nitrogen-rich formula with lower quantities of phosphorus and potassium. Once your buds begin to form—about 2 or 3 weeks in—you'll want to switch to a minimal amount of nitrogen, a moderate amount of potassium, and a higher amount of phosphorus. Phosphorus is linked with resin and seed production, making it ideal for this stage.

Throughout flowering, you can also add various supplements to your nutrients to enhance desired attributes of the strains you are growing. Shooting powders, carb-loaded products, and aroma and flavor enhancers are just some of the various additives you can add to your nutrient regimen. Whatever you decide to use, refer to the manufacturer's directions, as the ingredients in these supplements vary greatly from product to product.

Enhancing Quality and Quantity

Aside from nutrients and environment, you can use some of the techniques from the vegetative stage (such as super cropping and LST) in the flowering stage to enhance your crop's quality and quantity. Once implemented, some of these techniques will need to be continued during your plants' entire life cycle. However, the rewards are well worth it.

Lighting in Bloom

For indoor growing, adjusting your lighting schedule can have a big impact on your harvest. While beginners usually stick to a 12/12 lighting schedule (12 hours on and 12 hours off), more advanced growers sometimes use a schedule that will save time and money while increasing their crop's output. The following is such a schedule for a strain requiring 8 weeks to flower.

Weeks 1 and 2: 12 hours on and 12 hours off

Weeks 3 and 4: 11 hours, 30 minutes on and 13 hours, 30 minutes off

Weeks 5 and 6: 11 hours on and 13 hours off

Weeks 7 and 8: 10 hours, 30 minutes on and 13 hours, 30 minutes off

Finishing Touches

In the final weeks of flowering, remove blemished, dying, and excessive fan leaves to make later trimming somewhat easier. During the final week or two, you'll also want to use a flushing agent—or just plain water—to remove all nutrients from the roots, resulting in a better-quality, better-tasting bud free of any chemical residuals left by the nutrients used to grow the plant.

Fertilizing at the Flowering Stage

Throughout their life cycle, cannabis plants will require nutrients to thrive. The balance of nitrogen and phosphorus, along with the schedule of application of the nutrients, will change as a plant matures from the vegetative stage to the flowering stage.

Changing the Mix of Nutrients

In the vegetative stage, plants thrive on nutrients high in nitrogen. However, the flowering stage requires lower concentrations of nitrogen and higher concentrations of phosphorus. Premixed fertilizers will indicate the respective amounts of nitrogen and phosphorus available for flowering plants; in general, however, a mix including 5 percent nitrogen, 25 percent phosphorus, and 9 percent potassium works well. Other nutrients for flowering can include simple, all-natural carbohydrates like molasses, which can act as a "sweetener" for the buds, enhancing their flavor.

How to Fertilize and How Often

Fertilizing at the flowering stage is accomplished with a liquid or powder mix that's introduced to your plants' grow medium so it will be absorbed by the roots. Some fertilizers are gentle enough to be used at every watering, while others should be used every other watering or even less. No matter the case, it's important that your concentrations of fertilizers not be too high. Follow the manufacturer's recommendations with regard to parts per million (ppm), as too much of a good thing is bad for plants, resulting in little to no nutrients being absorbed.

About 2 or 3 weeks prior to harvest, the nutrients should be reduced and can even be stopped, and your plants provided only with water. This will allow the roots to be flushed of salts and other minerals, resulting in fresher-tasting and smoother-smoking marijuana buds.

Foliar Feeding in the Flowering Stage

Foliar feeding is a technique of feeding your plants by applying liquid fertilizer directly to the leaves. It's a great way to provide your plants both the nutrients and water they need as a supplement to root feeding. However, as buds begin to mature and grow larger and denser, foliar feeding isn't necessarily recommended. Moisture accumulates in the dense flowers and can cause mold to occur deep inside the buds. Therefore, it's recommended to only mist or foliar feed your marijuana plants the first 3 or 4 weeks to avoid the facilitation of mold growth in your buds.

Pruning at the Flowering Stage

Pruning at the flowering stage can help a cannabis plant use its energy to flourish to its maximum potential. Pruning at this stage is limited and should be as nonintrusive as possible to prevent shocking the flowering plant.

Tools

Like at the vegetative stage, a clean, sharp pair of small pruning scissors works best for pruning.

When

After the plants have been in flower for about 2 weeks, you'll start to see small leaves appear at the nodes of most fan leaves, where the leaf stem intersects the branch.

Why

The goal of pruning during the flowering stage is to direct hormones to the tips of the plants' branches by removing potential flowering sites nearer to the main stem. You also have the option of removing fan leaves during this stage in order to save yourself time during later trimming.

"Popcorn buds"—which only get about as big as a piece of popcorn, hence the name—are very small flowers that naturally occur throughout the plant at the nodes of leaves and stems. In addition to being small, the buds are generally underdeveloped and airy, due to the fact that these sites are competing for the same energy as the naturally larger buds near the ends of the branches and are receiving less light. Pruning these bud sites allows you to concentrate hormones where you want them—the largest buds on the tips of the branches.

In terms of removing fan leaves, once the leaves start to darken, hang downward, or blemish, simply assist your plants by removing them. This will allow the light to penetrate more deeply into the canopy and leave you with less to trim away after harvest.

Popcorn bud

How

You can remove the popcorn buds by either cutting them off with your pruning scissors or pinching them off with your fingers, being careful not to damage the surrounding plant tissue.

Use pruning scissors to gently remove popcorn buds.

The Last Pruning?

This may be the last time you'll need to prune. However, you may look at the plants after 4 weeks in the flowering stage to determine if another round of light pruning is necessary.

Keeping Flowering Plants **Healthy**

Once your plants are in flower, you're nearing the finish line. Although your plants are seemingly much more resilient at this point, they still need the attention and dedication you've provided them thus far. The following are some considerations for keeping your plants healthy during the flowering stage.

Foliar Spraying

As discussed earlier for keeping vegetative plants healthy, foliar spraying via fertilizers, pesticides, and water is a great way to supplement your garden. Foliar spraying helps your plants by aiding in nutrient intake, protecting them from pests, and providing moisture.

You won't be able to use foliar fertilizers all through the flowering cycle due to the increased risk of mold from too much moisture. Therefore, make sure you get this accomplished early on in the flowering stage. Most supplemental nutrient foliar sprays can be used up to 3 weeks into the flowering stage.

The safest foliar spraying pesticide products to use during the flowering stage (or any stage really) are Organic Materials Research Institute (OMRI) certified. However, because they're composed of organic materials, they're not the most potent should you experience a serious infestation. That being said, it would be wise to maintain your preventative pest maintenance during the first few weeks of flowering.

Adding water will increase your humidity levels and should only be used if your RH needs to go up a bit. Otherwise, you risk the chance of mold and mildew with unnecessary moisture.

External Device That Promotes Healthy Plants

The temperature of your flowering plants should be kept between 72°F and 78°F (22°C and 26°C), with some sativa strains able to go as high as 85°F (29°C). During this time, you want to keep the humidity levels lower than they were in the vegetative stage. For this reason, you need to invest in a dehumidifier, which removes excess humidity from the air. While the humidity levels are strain specific, you probably never want to go below 45 percent or beyond 60 percent relative humidity (RH). An RH of 50 percent is just fine throughout. However, if you decide to keep your plants at 55 to 60 percent RH, lower the humidity the last few weeks of flower to reduce moisture build-up in the buds.

Using CO_2 and Flushing

CO_2 of 1,200 parts per million (ppm) or less should be introduced during the flowering stage to assist in your plants' development and overall success. While delivery systems are discussed more in relation to indoor growing in Part 3, just know the application of CO_2 will help your plants use more light in order to grow faster.

Also, flushing your plants weekly or at least biweekly with a flushing agent allows for a much better nutrient uptake, which translates to a healthier plant in bloom. Once your crop is ready—about the last week or two of flowering—you'll want to begin an end-of-harvest flushing in place of the nutrient feeding.

Don't Transplant

Transplanting your plants isn't recommended in the flowering stage, as the stress recovery time will heavily reduce their output. Therefore, plan your transplanting accordingly so it happens prior to the flowering stage.

Flowering rooms without a dehumidifier are much more prone to undesirable issues, such as powdery mildew.

Harvest Stage

This is what it's all about! Finally, your work has culminated in the harvesting of your garden. Knowing when to harvest is very important, and harvesting times will vary strain by strain. However, let's take a look at the basics before going into the specifics of how to harvest.

The Size of Your Harvest

Depending on the size of your crop, you may want to take it down in sections to keep the buds from drying too quickly, which diminishes the quality.

When It's Time to Harvest

Once you've grown a particular strain a few times, you'll know when it's ready for harvest. However, during the first couple runs, you'll need to determine when it's ready according to the trichome development on your buds.

If you recall, trichomes are the resin glands that make the buds sticky and tend to have the majority of THC. They start off as a clear pinhead-shape protrusion before gradually turning opaque and then finally turning amber. Based on your preference, you'll harvest during the opaque or amber stage. While trichomes are best observed under magnification, some experienced growers are able to assess what stage of development they're at with the naked eye.

Though trichomes are really the only source you need to determine it, the state of the fan leaves can also indicate harvest time. When the life cycle of your crop begins to wind down, they will begin to show signs similar to that of leaves transitioning into the fall season, turning yellow or brown and beginning to fall off.

Dry or Trim First?

Before you harvest, it's important to decide whether you're going to dry and then trim your plants or vice versa. While some growers like to trim while wet to save time in the drying process, other growers dry the whole plant first to retain the flavor and smoothness of the buds.

If you decide to dry first, take your plants down and hang them in a room with 45 percent to 55 percent relative humidity (RH) and a temperature below 75°F (24°C)—but not too cold, or the plants will take too long to dry. Light air movement is ideal, and the implementation of a dehumidifier, humidifier, or both will ensure you stay in the "sweet zone" for the proper drying of your crop. The odor will be very prevalent at this stage, so make sure you have some type of odor control. One way to control odor is with fans that pull air through charcoal filters, basically scrubbing the air. Too much is never enough when it comes to controlling the smell during harvest, so prepare accordingly.

Trichomes turning amber.

Preharvest Tip

Throughout the flowering stage, you'll have the opportunity to remove dying or blemished fan leaves. Removing these leaves will lighten the load when you begin to trim after harvesting. Also, getting rid of these leaves prior to drying is essential in maintaining the buds' aroma. Too many leaves make your flowers smell like hay, and that's most definitely not what you're striving for.

How to **Harvest**

When the buds are mature, it's time to harvest the flowers from your plants. You can then reap your bounty and enjoy the peaceful and rewarding benefits of your labor.

Determining When to Harvest

A bud can gain up to 20 percent of its total weight during the last 2 or 3 weeks of the flowering stage, which is usually between 6 and 10 weeks into a flowering cycle, depending on your strains. Bushier plants will generally have a maturation rate in the 8- to 10-week range, while taller and lankier plants will generally have a maturation rate in the 10- to 12-week (or longer) range.

However, the best and easiest indicator of when you should harvest is the color of your trichomes:

Clear trichomes. These are trichomes in the first stage of maturity. Clear trichomes have a lighter effect, as they contain only precursor cannabinoids, which aren't psychoactive. Therefore, you'll likely want to wait until they are more mature before harvesting.

Opaque trichomes. These indicate the average "ideal" of maturity, as these contain fully realized THC. While it's impossible to harvest all of your plants when they have opaque trichomes, harvesting as many as possible when they are this color will yield some great marijuana.

Amber trichomes. These indicate the buds are definitely ripe. Amber trichomes contain a degraded THC, cannabinol (CBN), which is 90 percent less potent and produces less of a true high. Even though it's best to harvest before this time, amber trichomes produce more of a high than clear ones, meaning they'll be more potent later rather than earlier.

MATURE TRICHOMES
The beautiful opaque trichomes are starting to turn amber, indicating the plant is at the ideal stage for harvest.

Preparing for Harvest Time

Once you've determined when to harvest, you should make certain preparations before that time to ensure success. As you'll learn later, having a good harvest will make drying and curing your buds easier, which will ultimately influence the quality of your cannabis.

Over the course of 2 or 3 weeks prior to harvest, you should flush your plants, giving them nothing but fresh water. This helps get rid of the salts and minerals left by the fertilizers in the plants. Finally, a few days before harvesting the buds, you can allow your plants to droop just a bit, in need of watering. Less water for your plants means less moisture when it comes time to dry your harvest, saving time and effort.

When ready to harvest, be sure to have clean pruning scissors or shears at the ready (you can also wear latex gloves, though they aren't required). A good way to ensure a sterilized cutting surface is to soak your instruments in rubbing alcohol or vodka beforehand.

After Harvest

Some growers like to trim plants when they're wet so less vegetation has to dry and cure. Meanwhile, other growers believe drying and curing whole plants allows the buds to retain more flavor and smoothness. The choice is yours.

Harvesting Method

Harvesting your buds is as simple as cutting the flowers off your plants; however, your approach to this can determine the overall quality of your yield.

1 Using a clean cutting instrument, such as pruning scissors or shears, cut off entire branches covered with buds. Smaller buds that remain farther down along the length of the branches can be trimmed off.

2 Remove any remaining fan or shade leaves prior to a final manicuring trim. Final trimming can occur when plants are wet or dry, so you can also complete that now, after harvesting, or wait until the buds are dry.

Trimming the Buds

Now that you've harvested your crop, it's time to trim the buds. This is another important step in preserving the quality of your work, as trimming incorrectly can cause beautiful flowers to smell like hay. You can either trim wet or dry buds with a trimming machine or by hand. I'll discuss both techniques here, as well as the advantages and disadvantages of each.

Using Trimming Machines

Trimming machines come in two types: dry trimmers and wet trimmers. Dry trimmers are a relatively new and evolving technology that allows you to trim the flowers once they've dried. Trimming once the buds are dry allows you to transition directly into curing after you trim. The downside to this is a lot of the trichomes will be knocked off, plus it's difficult to separate the trimmed matter, leaving you with very desirable trichomes mixed in with not-so-desirable plant matter.

Wet trimmers allow you to trim the buds while they're still wet, prior to drying. The downside to wet trimmers is the buds are sticky and leave a lot of the trichomes on the blades of the trimmer. You'll also have a mix of the remains of trichomes and plant matter that's nearly impossible to separate. Another disadvantage is if you trim the buds while wet, you'll never be able to properly rehydrate your buds should they dry too quickly due to removal of the stems.

Trimming by Hand

Hand trimming your harvest is the old-fashioned and still-preferred method of growers who are looking for quality over quantity. While it takes much more time to trim by hand, the overall look and quality of the finished product is noticeably different compared to machine-trimmed flowers. Unlike with trimming machines, buds trimmed by hand are unique in their looks because you're using your knowledge of the strain to get the maximum amount of trichomes, not simply a uniform product. You'll also have the ability to have all of the plant matter separated in a way that will allow maximum efficiency in the processing of your byproduct.

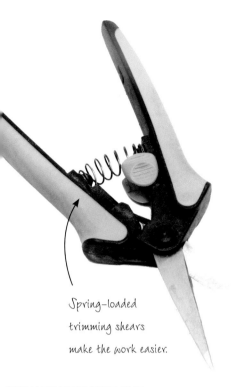

Spring-loaded trimming shears make the work easier.

TRIMMING EQUIPMENT

All you really need for hand trimming are some latex gloves or something similar and some good trimming shears. Spring-loaded trimming shears can be found at your local garden supply store and work much better than scissors.

HOW TO TRIM BY HAND

The hand trimming method can be broken down into three steps.

1 Cut your plants into manageable sections by trimming off all the side stems from the main stalk and removing the colas from the center stalks.

2 Cut off all the fan leaves. Remove the smaller leaves as well, trimming as close to the buds as possible without losing any flowers. While you should discard the fan leaves after removal, feel free to keep the smaller trichome-covered leaves for hash-making purposes.

3 Trim off anything else that's not the marijuana buds, being careful not to disturb the buds in any way.

Drying

The basic intent of drying is to remove much of the moisture and chlorophyll, improving the taste and smoothness of your product, without overdrying the buds. Prior to harvesting, you'll need to decide whether you want to dry before or after trimming. While some believe waiting to trim allows the buds to have a stronger aroma, it's really a personal preference. Don't jeopardize all your hard work because of poor planning! Whatever you decide, the following are a few things to help make the 10- to 14-day drying process a successful endeavor.

DRYING LINES
Once you've broken down your plants, use the branches as hooks to hang them on well-anchored drying lines.

Removing Fan Leaves

Removal of all the fan leaves prior to drying is a must. Whether you trim your plants as a whole before or after drying, this step will help in the overall taste, smell, and quality of your final product. Many growers make the mistake of simply trimming away the protruding fan leaves. However, you'll get a finer-quality product if you clip fan leaves where the stem meets the node.

Breaking Down Your Plants

Separating your plants into smaller sections will also help the drying process. Buds have a tendency to pull moisture from the stems, so having less stem means less moisture retention. If you try to separate your plants into too many sections, you may find they dry too quickly (based on your test to determine whether drying is complete), which can affect the quality. To avoid this, you have the option of doing the reverse and keeping your plants close by tying sections together. While the slight addition of moisture can rehydrate your flowers, this is an inexact procedure, and overdoing it can cause serious damage. You can also buy moisture packs to help overdried flowers become somewhat moist again, but only to a certain extent. Be patient, and you'll be much more satisfied with the outcome.

Controlling the Climate

You'll want your room or drying area to be dark, if possible, with about 50 percent relative humidity (RH) and a temperature around 72°F to 75°F (22°C to 24°C). While this can vary slightly, any alterations will change the dry time. To have complete control over your drying process, you should install a humidifier or dehumidifier—and in some instances, both, to deal with moisture fluctuations—depending on your local climate. You may also want to use air conditioning to keep the room from getting too warm. Some slight air movement helps as well to regulate your climate.

Determining When Drying Is Complete

Once your flowers have been drying for a few days (before the 10- to 14-day finish time), begin to bend the stems and see if they snap. Ideally, you want the snap to occur in the 10- to 14-day period. You may find your plants dry faster if they're in too-small sections, as discussed previously, or if you trimmed them wet. If you don't want this to happen, implement the measures described previously so you can control the climate and the subsequent outcome.

DRYING PRIOR TO TRIMMING
This plant has been dried prior to trimming. Because the leaves curl up around the buds, it can make it more difficult to remove all of the leaves from the buds. This is why some growers prefer to trim prior to drying.

Curing and Storage of Cured Pot

One of the most important and greatly overlooked stages in the cultivation of cannabis is the curing process. If your plants aren't cured properly, you'll be horribly disappointed; after all, you don't want the effort you put into growing your crop to go to waste. You want the odor, stickiness, potency, and taste of your cannabis to be solidified during the curing process.

When to Start Curing

After your buds have dried, you should remove them from the stems and ensure no unwanted plant matter is present. Once you've done this, you can begin the curing process.

> ### **Drying Note**
> Remember, the stems should snap when bent before you start removing the buds from them. This is the main indicator that the buds are dry enough to begin the curing process.

Glass mason jars are the best containers used to cure.

Open the lid just slightly to "burp" the buds of gases.

Storing Your Pot During and After Curing

Glass containers, such as glass canning jars, work very well for curing. Plastic containers or oven bags can sometimes negatively affect the overall quality of the cure. And the THC trichomes cannot be removed from plastic containers as easily as they can from glass. You basically want containers that allow you to have control over the airflow into and out of them. To allow that airflow process, fill the containers only 75 to 85 percent full of buds, or to the point the buds can still move around when you shake them. Store your containers in a cool, dark place because heat causes the pot to dry out and light harms trichomes. Humidity-control products are available that help maintain the perfect moisture content of the cured marijuana.

Burping

During the curing process, you'll "burp" the containers. This simply means you open the containers to release gases from them. Burping should be done a few times a day during the first several days, and then tapered off throughout the next couple of weeks as needed. After the first month or so, you should only need to burp your jar about once a week or so. When burping, shake the buds around in the container to ensure a total mixture of air and the release of water vapor. You'll notice a change in the odor throughout the burping process and see your work of art finally come to fruition.

How Long to Cure

Your cannabis should be properly cured in about 1 month. By that time, the buds should feel a bit sticky to your fingers and move independently, not clump together in bunches, when you shake the container. However, some say you can continue to cure cannabis for up to 6 months to increase its potency. Humidity-control products can both extract excessive moisture and add moisture if needed to maintain a perfect cure.

Making **Kief**

Basically, *kief* refers to the trichomes that contain the terpenes and cannabinoids that make cannabis what it is. Making kief is basically the precursor to processing trichomes into hash.

Passively Collecting Kief

Kief is easily collected from very dry marijuana buds. The more trichome coverage the bud has, the more kief you'll be able to collect.

A common passive method of collecting various grades of kief is storing your marijuana in so-called kief or personal-use storage or stash boxes, which allow you to gather kief through the natural denigration of vegetative matter as it's handled during regular use. Various screens filter various grades of kief in the bottom of these boxes.

Kief can also be prevalent in the bottom of whatever storage container you use for your personal-use cannabis, whether it's the bottom of a small bag or jar. However, I highly recommend you store your usable marijuana in glass—as resins tend to stick or bond to plastic—and screen it of plant matter before use.

Actively Collecting Kief

While the passive methods of collection only yield a small amount of kief of varying quality, actively sifting your buds gives you both a higher amount and a finer quality of kief. Mesh screens of around 150 microns will allow the largest amount of trichomes to be collected while sifting out most of the other plant matter.

WHAT YOU NEED
Frame
1 × 1-foot (30.5×30.5cm)150-micron mesh screen
Heavy-duty staple gun with staples
Tray or bag large enough to fit under screen
Dried buds

1 Stretch the mesh and secure it around the frame with staples. Place the frame above a tray or bag. Remove the nonresinous leaves and other vegetation from the buds by hand.

2 Gently rub the buds into the mesh screen. Resins will fall through the mesh into the collection tray or bag as a fine powder.

Hand Grinders

Kief can also be efficiently collected in the screens of some hand grinders used for crushing marijuana buds into a consistency suitable for rolling a marijuana cigarette, also known as a *joint*. Just be sure your hand grinder has one or more screens for kief collection.

Screening for Quality

Kief can be separated into different grades of quality through the screening process. Different screens sized by microns filter particulates such as plant matter and trichomes by size, allowing them to be separated. The smaller the microns of the screen, the better the grade of the kief. Generally, a 100- to 150-micron screen provides about the right size for the most trichome collection. One thing to consider when "kiefing" your buds is that removing trichomes likely will reduce the potency of the flower itself.

Making **Dry Hash**

Once you've extracted kief, you have the material to make dry hashish, or hash. Hash is a great delivery system for the active ingredient in cannabis, THC. With the trichomes of your buds collected via the kief process, you can create a time-tested, tried-and-true, potent concentration of this form of marijuana using the following method.

A Brief History of Hash

While the records of cannabis use go back thousands of years BCE, the records of widespread use of hash originate in Arabia around 900 CE. Early in the twelfth century, hash smoking became very popular in the Middle East. By the late seventeenth century, hash had become a major trading commodity between Central Asia and South Asia. Hash later rose in popularity across Europe and America during the 1960s. From the 1960s through the 1980s, Morocco was the largest producer and exporter of hash in the world.

WHAT YOU NEED
Strainer
Cellophane
Tape
Parchment paper
  (Newspaper can be
  substituted in a pinch but
  is not recommended.)
Spatula
Baking sheet
Rolling pin

1 Preheat your oven to 350°F (180°C). Collect trichomes using the process described in making kief and strain through a strainer.

2 Collect strained trichomes and place them on the cellophane, packing them in as you do so.

3 Roll up the cellophane as tightly as possible and tape so the block holds together. Wrap in an outside layer of parchment paper and tape that as well.

4 Place the covered block on a spatula and, being careful not to burn yourself, pour hot water over the block until it's saturated. Place the block on a baking sheet and put in the oven for 10 minutes.

5 Using the spatula, place the block on a countertop, cutting board, or other hard surface. With a rolling pin, press the block to thin it, applying pressure as evenly as possible. (This may take only one press of the rolling pin.)

6 Place in the refrigerator to cool. Once cool, peel away the paper and cellophane, and enjoy!

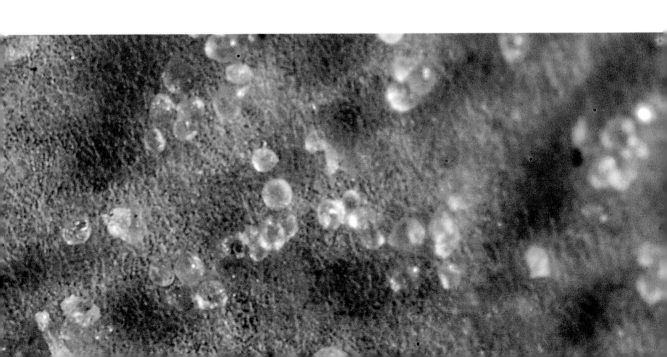

Making Water or **Bubble Hash**

Using very cold water and ice, you can make a clean and concentrated form of hash known as *water* or *bubble hash*. The potency and pure essence of the plant that comes through in the final product makes it a well-liked form of hash.

Bubble Hash

If you're planning to make bubble hash, freeze your usable cannabis trimmings or flower immediately after harvest. Keep frozen until use.

The Difference Between Water and Bubble Hash

The methods for mixing and processing water and bubble hash are virtually identical. Both involve mixing marijuana with water and ice and then straining the mixture through different micron-sized bags. However, unlike water hash, bubble hash—hash that can actually bubble or melt when heat or an open flame is applied—is made from marijuana buds that have been frozen prior to drying, immediately after harvest.

WHAT YOU NEED

You'll need 1 ounce (25g) of marijuana buds or trim per 1 gallon (3.75 liters) of water. Beyond your marijuana, you'll need the following equipment.

Bubble bags. In sizes from 1 gallon (3.75 liter) to 30 gallons (113.5 liter), these bags strain the mixture in different stages, allowing trichome resins to be kept on some layers while trapping other plant matter on other layers. You'll be layering four bubble bags of different microns for this process: 25, 70, 160, and 220. You can find bubble bags online or at most hydroponic supply stores.

Bucket or washing machine. This is where you'll place the bubble bags and mix up your marijuana. The cheapest way is to use a 1-gallon (3.75-liter) bucket and mix by hand; however, you can invest in a small washing machine (like those found in recreational vehicles) to mix larger amounts.

Mesh laundry bag (optional). If you don't want to have your cannabis loose during the process, you can place it in a mesh bag. This type of bag can be found online or at any big-box store.

Ice-cold water. This is used to make the trichomes stand up on the cannabis, making it easier for them to fall off the leaves.

Ice. Like with the water, the ice makes it easier for the trichomes to fall off the leaves. You'll need ½ to ¾ of your bucket or machine to be filled with ice.

Cardboard or parchment paper. As the final part of the process, your hash is spread on cardboard or parchment paper to dry.

Method

MIXING

The first part of making water or bubble hash is mixing your marijuana with water and ice.

1 In the bucket or washing machine, place the bubble hash bags one inside the other sequentially, starting with the 25-micron bag and ending with the 220-micron bag. The 220-micron bag is considered your working bag where you'll add ice, water, and your marijuana.

2 Fill the bucket or washing machine with ice until it's approximately half full.

3 Place the usable cannabis trim or broken-up buds into the mesh laundry bag, and then put the bag in the bucket or washing machine. (Alternatively, the marijuana can be added loose.)

4 Add ice-cold water until the top layer of ice just starts to float.

5 Top off with more ice.

6 Mix with a gentle motion, either by hand or by machine, slowly and thoroughly for about 15 to 20 minutes.

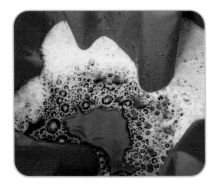

7 When mixed properly, small bubbles of air will make their way to the top of the ice and water, resulting in a frothy, foamlike consistency. Allow the mixture to settle for approximately 15 minutes.

Mixing Caution

Be careful not to whip or blend the mixture to the point that it grinds up or breaks apart too much plant matter. The primary goal is to completely wash the frozen trichomes from the usable marijuana, separating it from the other plant matter.

PROCESSING

The screens in the bottom of the bubble bags allow water to drain from the mixture, separating the hash from other plant matter. The bags strain the mixture in different stages, allowing trichome resins to be kept on some layers while trapping other plant matter on other layers.

1 Drain the water from each successive bubble bag, leaving the hash resins on the screens at the bottom of the bags.

2 Scrape the resins from the screens of each of the bags.

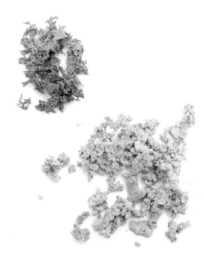

HASH QUALITY
The quality or grade of hash is different based on the size of the micron screens.

DRYING

Drying and curing water or bubble hash should be done properly to ensure a high-quality smoking experience. This is accomplished by drying the hash on parchment paper or cardboard in a ventilated but not breezy environment.

1 Spread the hash on parchment paper or cardboard in as thin a layer as possible.

2 Allow the hash to dry in a ventilated but not breezy place (such as a dark, dry cabinet) for 2 or 3 days.

Drying Note

The trick to properly drying and curing water or bubble hash is to do it slowly enough that it doesn't dry too fast or take too long to dry. Drying too fast can result in an overly dehydrated product that delivers a harsher smoking experience. Conversely, allowing the hash to remain moist for too long can facilitate and encourage mold growth.

Growing **Indoors, Outdoors,** or in **Grow Structures**

Deciding where to grow your marijuana for personal use is a logistical one, dependent upon both environmental factors and personal choices. Variables you should consider when making the decision include convenience, climate, and the laws governing cultivation of cannabis for personal use in your state.

Why Grow Indoors?

Until the recent reforms of marijuana laws, the prohibition of marijuana has demanded that cultivation occur covertly indoors. Over time, skills and technologies have been refined to allow for near-ideal controls of climate that allow you to harvest great crops indoors. So while it's not the only option anymore, indoor growing is still a popular way to go.

When growing indoors, you must provide an adequate year-round climate with regulated airflow, humidity, temperature, pest control, and lighting. Because care must be taken to maintain these conditions to avoid damage to your plants and your property, the initial investment for indoor growing makes it the most expensive option to consider. However, if you want more control over the elements and don't have the space for a larger grow, you should consider growing indoors.

Why Grow Outdoors?

Growing outdoors can be very pleasurable and productive. Most of the conditions that would need to be virtually re-created and maintained in an indoor environment can oftentimes be found already in perfect balance in nature. Plus, some growers find growing in soil to be the ultimate, "beyond-organic" experience due to it being environmentally sustainable, all-natural, and often pesticide-free.

The most important consideration for outdoor growing is the climate in which you plan on growing. Most of North America provides ideal climates for growing outdoors; however, some areas have shorter growing seasons or wetter climates, which may not be as conducive to completely open-air growing of cannabis.

Why Grow in Grow Structures?

Grow structures can provide the perfect option between the extremes of indoor growing and outdoor growing. Grow structures combine the benefits of both an outdoor grow with some of the controls of an indoor grow, such as the following:

- Growing seasons can be extended using both natural and artificial lighting.

- Plants can be protected from extreme weather.

- Grow structures can be sealed to allow for humidity and temperature controls or opened to allow for natural airflow.

Grow structures come in all shapes and sizes and can be constructed as inexpensively or expensively as you can imagine, giving you a level of control over your investment.

Part 3
Indoor Growing

Indoor growing is a preferred method for those who have the space and time to do so and are looking for quality over quantity. The major advantage of an indoor grow is that every day can be a perfect day in a climate-controlled, secure, pest-free environment.

Introduction to **Indoor Growing**

Imagine if every day were a perfect day—the sun was just right, the wind blew perfectly, and the perfect amount of humidity was in the air. Consider having the ability to keep unwanted pests from entering your garden and to make sure your garden has every possible resource it could ever need. This is the beauty of the indoor grow, and when set up properly, an indoor grow is basically a cultivation lab for a healthy crop. However, you need to make certain considerations when deciding whether to grow your marijuana plants indoors.

What Plants Need Indoors

The key to growing indoors is to convince the plants they're actually outdoors. This means your indoor grow area will consume a lot of power, as the cost of operating lights—and the air conditioning required to cool those lights—are substantial when compared to an outdoor grow. Air (oxygen and CO_2), water, nutrients, and the right temperatures can also trick your plants into thinking they're living the perfect life every day. If you feel you can invest in the tools necessary for these factors to successfully pull this off, your plants will thrive indoors, never knowing the difference.

Soil vs. Hydroponics

A great benefit of growing indoors is the option of growing in hydroponics or soil. While most indoor growers prefer to grow hydroponically, both options are viable for your crop.

Growing in hydroponics produces quicker turnarounds but takes quite a bit of dedication in keeping everything as clean as it should be. This might make you wonder if you're a gardener or a maintenance person at times, and could ultimately cause more frustration than it's worth. While soil is very forgiving and definitely a better choice for beginners, it comes with its own set of obstacles as well. Choose wisely which direction you go, as the success of your garden ultimately lies in your decision-making and subsequent work ethic.

Making the Choice

Throughout this book, I discuss the various pros and cons of each medium and grow method in detail. Beyond that, research what will work best for you and the space in which you're cultivating.

Choosing Strains for Growing Indoors

Probably the most beneficial reason to grow indoors is the fact that you can control the environment and seasons. This means you can harvest all year round in a perfect environment. That being said, choosing the right strains to grow indoors is very important and can maximize your productivity. Most indoor growers prefer to cultivate indicas because they require less time in bloom, but some hybrids and even some sativas also do well indoors.

Space Considerations

The space in which you'll be growing should account for a major part of your decision-making when it comes to strains. If you have unlimited space to grow, just about any strain will work. However, many growers don't have the luxury or desire to "go big." For instance, low ceilings mean you'll want a strain that can be trained to work in that environment or that naturally grows like a vine (such as most OG varieties).

If you're growing in a hydroponic system, you'll have different considerations than if growing in soil. Look for hardier strains, such as any Kush or indicas.

Level of Experience

Your level of experience with growing can also influence which strains you choose. If you have quite a bit of experience, the options are pretty much only limited to your space. However, if you're new to growing marijuana, you'll likely want an auto-flowering strain. Auto-flowering strains automatically switch from the vegetative stage to the flowering stage with age rather than variation in light and dark hours, making them more forgiving. Northern Lights Auto and White Widow Auto are a couple examples of auto-flowering strains commonly recommended to beginning growers.

Popular Strains for Growing Indoors

Even after narrowing down your choices by space and ease of growth, you'll still have hundreds of strains to choose from. If you're unsure where to start, the following are some popular strains cultivated by marijuana growers indoors.

- **Afghan Kush.** This indica strain is an easy-to-grow weed with dependable yields. It has a potent hit with a heavy, sweet aroma.

- **B-52.** Popular with beginner and experienced growers alike, this hybrid grows prolifically and produces an energetic, joyful high.

- **Blue Dream.** This sativa-dominant strain is highly potent and has a high yield. It provides a quick cerebral rush, followed by a relaxing body high.

- **Granddaddy Purple.** This indica strain grows short and squat, with many side stems. It produces an energetic high and has a distinctive fruity smell.

- **OG Kush.** This popular indica strain favors an indoor setup for growth, preferably using hydroponics. It has an earthy, musty flavor and a prolonged high.

- **Ultra Lemon Haze.** This hybrid is an auto-flowering strain that grows well indoors. Smokers can enjoy an energetic high from this plant.

INDICA STRAIN
Indica strains are great for growing indoors because the plants produced are short in size.

SATIVA STRAIN
Sativa strains are better for outdoor growing, but you can grow them indoors using methods to control their height (such as low-stress training).

Grow **Mediums**

When growing indoors, your options for grow mediums increase substantially. This is mainly due to the fact that most indoor growers implement a hydroponic growing method, where various mediums can be utilized. Let's take a look at the various grow mediums you can use, as well as the pros and cons of each.

Soil

You can grow your plants with many types of soil, from store-bought mixes to ones you put together from scratch. Whether you purchase a premixed soil or make your own, a soil should have a good composition of nutrients in order to help plants flourish. Soil is more forgiving than most other mediums because it retains moisture longer and forces the roots of your plants to work. And if you decide to have your garden in individual soil containers, soil allows you to more easily transport your crop.

One downside to soil is it's an ideal breeding ground for many unwanted pests that could damage your plants. Another problem with using soil is disposing of it when it's finished the growing cycle. This can be mitigated by reusing your soil or sending it to landfills. If you have the time and space to amend your soil (by adding nutrients to adjust the pH, fix the consistency, and so on to make the soil viable for growth again), you'll save a ton of money; however, amending your soil incorrectly can cause more trouble than it's worth.

Rock Wool Cubes

These cubes are actually spun from heated rock in a way similar to wool, hence their name. Rock wool cubes are used in most commercial grow operations and are fed through a drip stake at timed intervals in hydroponic systems. A versatile medium, rock wool cubes can be purchased in many different sizes. Plus, they're trustworthy for cloning, as you won't be dealing with fragile roots during transplant, which results in faster growth of stronger roots.

A downside to rock wool cubes is their propensity to grow algae. Also, if your feeding schedule is off, this medium can retain more water than intended, causing root rot and other debilitating issues.

This is a prefertilized soil. The white pebbles you see are perlite, which is used to help aerate the soil.

Coco Coir

Coco coir, a natural fiber extracted from the husk of coconuts, makes a great grow medium. It works similar to soil but dries faster, allowing for more frequent feedings.

However, gnats have a tendency to show up when you use coco coir, so have a plan in place if you decide to make this your grow medium.

Expanded Clay Pellets (Hydroton)

Expanded clay pellets (known by the brand name hydroton) can either be used as a primary or supplemental medium to help aerate soil or coco coir. Hydroton is mainly used in hydroponic applications, as the clay pebbles are inert and retain moisture, yet still provide plenty of surface area for air molecules. When used as a supplemental medium, it can either be placed at the bottom of a pot with soil above or mixed directly into the soil. Either of these methods will allow the soil to breathe and in turn allow the roots to dry faster in between feedings.

Because hydroton retains water, a failure in the hydroponic system can in turn cause major issues for the grow medium. For instance, if a pump fails, your plants will die quickly due to the lack of moisture.

Coco coir is much lighter and airier than soil, allowing it to dry faster.

Setting Up a **Grow Room**

When setting up an indoor grow room, you have many decisions to make, most of which are based on cost, where you live, and the legality of the operation. The intended size of your indoor garden is also a major consideration, as the larger the grow, the more resources you'll need to power and sustain your operation.

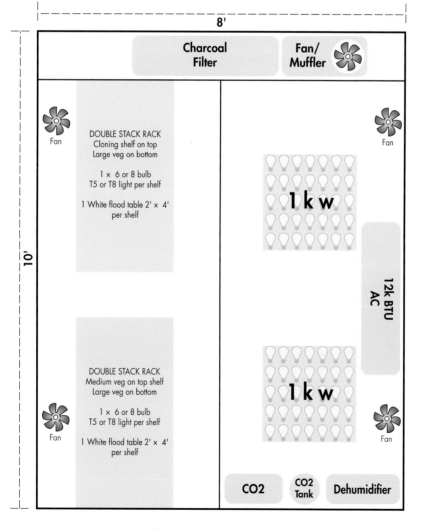

BASIC GROW ROOM SETUP

Here is a 10 × 8-foot (3×2.5m) grow area representing all stages of growth. The room has been divided so individual control can be achieved between the vegetative and flowering areas.

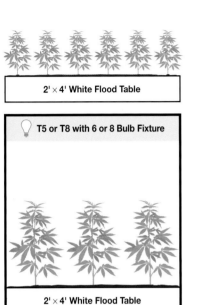

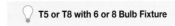

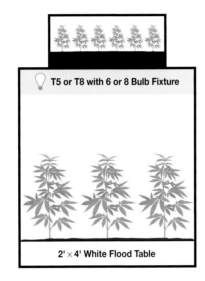

SIDE VIEW

This is a side view of the rack configurations pictured in the grow room diagram to the left, which includes flood tables and lighting for your plants.

Determining Your Budget

Probably the biggest determining factor in the planning of most projects is your budget. We often have desires bigger than our wallets, but this is one scenario you want to be sure not to undervalue. Certain factors can influence your grow room's budget:

- The medium in which you will be growing: soil, coco fiber, or hydroponics

- Whether it's a closed-loop operation, where you'll be cloning your own plants or possibly starting all from seed, or a flower-only operation, where you'll procure new crops from an outside source

- The type of lighting you'll be using

Once you decide on these factors, you can begin to formulate a budget based on your needs.

Checking the Legality

Before making any financial commitments, check the particular laws of your state or country in order to see what sort of restrictions and regulations there are on cultivating. You can do so at NORML.org/laws.

Indoor Grow Tents

Indoor grow tents provide a great alternative to a full-size grow room and offer a proven track record of success for a small, self-contained, and relatively unobtrusive indoor growing option in almost any space.

Selecting a Growing Method

To start the planning process, select a method with which you plan to grow your plants. While these are explained in more detail throughout the book, your basic methods of cultivation will either be soil or hydroponic. This decision is affected by whether you intend to water your plants manually (soil) or through an automated watering system (hydroponic). While it may seem like an automatic watering system would be ideal, hydroponic systems require some maintenance and a lot of cleaning. So weigh this choice carefully as you also think about how you're going to water your garden.

Choosing a Grow Medium

After you select the method, you can choose the medium in which you grow your plants. If you're a beginner, it's probably best to start with soil, as it's the most forgiving of the mediums. If you feel a bit more advanced, you can grow with coco coir. Unlike aeroponic or hydroponic applications—which keep the roots of plants drier prior to watering—soil and coco coir will provide you with a larger margin for error, as they will remain moister longer. In hydroponic systems, you have the options of rock wool (which isn't as forgiving as other mediums) and clay pellets.

Selecting Lighting

After choosing what you'll grow your plants in, you need to decide what type of lights you'll be using in your indoor garden. The lighting required in the vegetative stage will be different from those in the flowering stage, and both will create different climates. A safe bet is to go with florescent bulbs in veg and HPS bulbs in bloom. (I go into more detail in the lighting, bulbs or LED lighting, ballasts, and reflectors sections of this book.)

Environmental Requirements

Each stage of growth will have different requirements for optimal results:

- Clones will need a warmer environment that's very humid at first, which is usually accomplished using a humidity dome and tray.

- The vegetative stage will be less humid but still around 55 to 75 percent, depending on the strain and age of your plants.

- The flowering stage should be at 50 to 60 percent humidity, depending on the strain.

- The drying process should be done with the humidity between 50 and 65 percent.

You can control the humidity levels of clones with a humidity dome and tray, while the humidity of plants in the vegetative and flowering stages can be controlled by way of a humidifier or a dehumidifier. Your ideal temperature in all stages should remain around 72°F to 80°F (22°C to 27°C) when the lights are on and not drop more than 20 degrees when the lights are off.

Going Green

You can still have a green mentality even in an indoor garden. For instance, disposing of your nutrients—diluted with water by 50 percent—in a garden or lawn application rather than dumping them into the sewer is much friendlier to the environment.

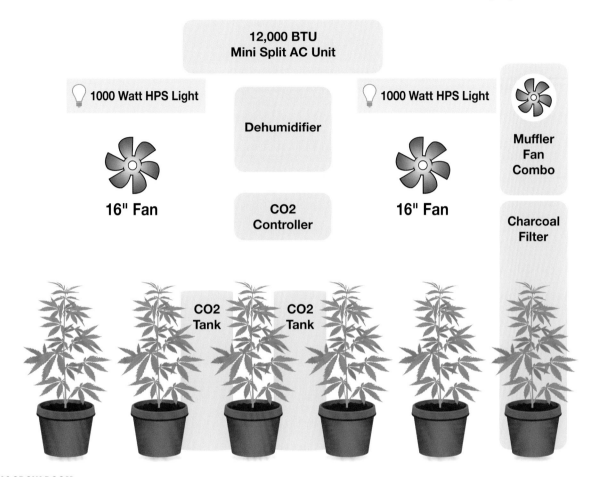

12,000 BTU Mini Split AC Unit

1000 Watt HPS Light

Dehumidifier

1000 Watt HPS Light

Muffler Fan Combo

16" Fan

CO2 Controller

16" Fan

Charcoal Filter

CO2 Tank

CO2 Tank

BASIC GROW ROOM

Here is a basic grow room layout with all the equipment needed to have a successful crop, such as fans, lights, CO_2 emitters, and so on.

Air Considerations

Air conditioning, fans, air scrubbers, and the addition of CO_2 will control your air. This equipment can be quite costly, but in the long run, you'll be glad you purchased all of it.

When it comes to air conditioners, you'll need 4,000 British thermal units (BTU) for every 1,000 watts of light in your grow room. As far as fans, you want moderate air movement to mix and circulate the air, as well as help your plants grow stronger. You can also use air

scrubbers to move the air or simply to remove odor from your grow room, though they may need to be outfitted with mufflers to reduce any unwanted sound. Finally, if possible, you should introduce CO_2 into the air, as it allows your plants to grow faster. The process of introducing CO_2 can be accomplished in several ways, which I'll describe in more detail later in this book.

Creating the **Ideal Indoor Climate**

In addition to choosing strains, grow mediums, and the proper grow room setup, the environment in which you grow your garden is just as important. The humidity, CO_2, air movement, and temperature will play as big of a part in the success of your crop as anything else. Of course, these ideal settings will vary according to strain. The following are the ideal settings for your plants in the various stages of growth and how to achieve them.

Germination Climate

If you're starting with seeds, you want to place them in a warm, dark area and keep them very moist at all times. This can easily be achieved by placing the seeds on a plate, wrapping the plate in a damp paper towel, and storing the wrapped plate in a closet or any undisturbed, dark place. Ensure the seeds stay moist and don't dry out; otherwise, you'll have to start over.

Cloning Climate

When cloning, you want light on the cuts 18 to 24 hours and a temperature about 75°F to 85°F (24°C to 29°C). The warmth from the lights will keep the root zone warm enough for roots to develop. You'll also need high humidity for the first 3 days. One way to accomplish this is to spray the cuts frequently with water. Another option is to put them under a humidity dome, which allows light and heat to pass through while trapping moisture inside. If you decide to use a humidity dome, reduce the humidity level slightly each day by opening the vents, making sure not to let it go below 60 percent. No matter what you decide to use, you can test the relative humidity (RH) with a hygrometer, which can be purchased at most home improvement or garden stores.

Vegetative Climate

The climate most conducive to the vegetative state is similar to a rainforest, with higher humidity levels of 60 to 70 percent and a temperature between 75°F and 80°F (24°C and 27°C). Because you'll want air movement in the room so the plants' branches can grow strong, you'll need some oscillating fans. (However, don't overdo it, as too much air movement can cause windburn.) And while some growers use it in their indoor grow rooms, supplemental CO_2 isn't necessary at this stage.

HUMIDIFIER

A humidifer is ideal for cloning and the vegetative stage in your grow room, when you need high humidity in order for your plants to thrive.

MINI SPLIT AIR CONDITIONER (AC) UNIT
This unit recirculates the air in a room rather than having a return like regular split units have, making it ideal for a CGE.

FANS
Fans, like this one sticking through a ventilation pit, help keep plants cool and exchange the air in your indoor grow room.

Flowering Climate

Once your plants are in the flowering stage, the humidity levels should be dropped to around 50 to 55 percent and monitored closely, as too much moisture in bloom will definitely cause problems in terms of mold and mildew. As for the temperature, it should be 72°F to 78°F (22°C to 26°C) during this stage. You also introduce CO_2 to create the right air mixture for your plants. The ideal amount of CO_2 in your bloom room is 1,200 to 1,500 parts per million (ppm) and can be generated through CO_2 tanks and a regulator. Either of these methods will need a monitor/controller to ensure your CO_2 levels don't reach deadly levels. The ambient CO_2 in the air is generally around 360 ppm, and if you're growing in a closed growing environment (CGE) like an indoor grow room, you'll definitely need to add CO_2 so your plants can take up enough nutrients to thrive.

Ventilation

When growing your plants, you want to ensure they stay at the correct temperature and have a steady supply of CO_2 as necessary. This is where ventilation comes in. A ventilation system will keep your indoor grow room from becoming too hot and will properly exchange the air so your plants are getting fresh CO_2.

Just some of the fans you can use for proper ventilation are oscillating fans, squirrel fans, and extractor fans. The first two help keep plants cool by blowing air between them. As the name implies, the latter fan extracts or pulls hot air out of the room and replaces it with cool air, or exchanges air for fresh CO_2. These are minor investments that have a major impact on the health of your plants.

Lighting

The greatest difference between indoor and outdoor growing is the light source. That being said, it's obviously the intent to replicate the sun as much as possible through each stage of growth indoors. Fortunately, not all stages require intense light, and the implementation of a stage-specific lighting scheme can save lots of money. The following are the basic lighting components you'll need at different stages, as well as what kind of lighting schedule you should follow as a beginner.

Lighting During Cloning

Nothing more than simple florescent lighting is required for this stage, and to be honest, anything more could prove detrimental to your plants' development. Keep the light close to the clones to provide the source of heat necessary for their development.

Lighting During the Vegetative Stage

During the vegetative stage, your lighting options expand. Although florescent lights will work just fine here, more intense lighting (such as HID bulbs) will enhance your plants' development and prepare them through a process called *hardening*, which I'll discuss in detail a little later in this book. Some growers simply adjust the bloom lights for the vegetative stage and gradually raise the intensity as they transition into the flowering stage.

Lighting During the Flowering Stage

In the flowering stage, the lighting options can become overwhelming, with bulb types, ballasts, reflectors, and combinations thereof. For instance, growers may use HPS bulbs with digital ballasts and vented hoods. Placement of the lights is also a very important factor. The higher your lights, the bigger the footprint; however, this will also affect the light penetration. You're looking for a "sweet zone" for each lighting configuration—one in which the lumen outputs are most efficient for your plants. This sweet zone shouldn't burn your plants from being too close or stretch them from being too far away.

Creating a Lighting Schedule

Once you've decided on your lighting scheme, you'll need to decide on your lighting schedule. Beginners usually keep the lights on for 18 to 24 hours during the cloning and vegetative stages before adjusting to a 12/12 (in other words, 12 hours on and 12 hours off) light cycle during the flowering stage. However, as with everything else I've discussed regarding strain-specific conditions, researching the environment in which your strain grows naturally can help you replicate its natural light cycle.

How to Measure Your Lighting Output

Photosynthetic active radiation (*PAR*) is measured in micromoles and is the most pertinent measurement of light when it comes to growing. Desired PAR should range between 400 and 700 in micromoles and can be measured with a PAR meter (which you can buy online). Another frequently mentioned aspect of lighting output is *lumens*. A lumen is simply the measure of the amount of visible light from a light source, meaning it's more about how bright the light is rather than how useful it is in photosynthesis. *Kelvin* (*K*) pertains to the temperature of the bulb and doesn't really equate to the photosynthetic properties required for growing plants.

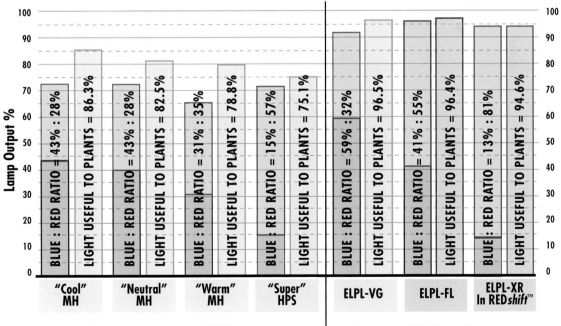

Blue:Red Ratios and Useful To Plants Light Output %

Lamp Output %

Lamp	Blue : Red Ratio	Light Useful to Plants
"Cool" MH	43% : 28%	86.3%
"Neutral" MH	43% : 28%	82.5%
"Warm" MH	31% : 35%	78.8%
"Super" HPS	15% : 57%	75.1%
ELPL-VG	59% : 32%	96.5%
ELPL-FL	41% : 55%	96.4%
ELPL-XR In REDshift™	13% : 81%	94.6%

High Intensity Discharge (HID) Lamps | **EconoLux ELPL Series Lamps**

This table shows the various spectrums of color that bulbs discharge in correlation to the percentage of useful light to plants.

Bulbs

Lighting is one of the most important factors in growing. No light means no growth, and poor lighting means your crop won't do very well. Each stage of growing will have different light requirements, with several options for each.

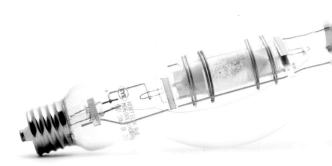

Fluorescent Bulbs

Fluorescent bulbs have been the standard choice for most growers due to their efficiency and low cost to run and replace. These bulbs can be used during cloning, which is the easiest stage to illuminate, as well as the vegetative stage. Several sizes and shapes of fluorescent bulbs are available, but most growers opt for a T5 or T8. The difference between T5s and T8s are size, lumen output, wattage consumption, and heat generation. T8s are less expensive and use less power, which translates to less heat generated. T5s have a higher lumen output, but the plant can't be as close to them due to the heat created by the higher-watt bulbs. You can get T5 high-output (HO) bulbs, too. These consume even more power and create more heat, but they ultimately produce a more intense white-colored light. The T5 and T8 fixtures now have LED-equivalent bulbs that replace the fluorescent T5s and T8s.

HID Bulbs

High-intensity discharge (HID) bulbs can be used in the vegetative or flowering stages. The two types of HID bulbs used in growing are metal halide (MH) and high-power sodium (HPS) bulbs.

MH BULBS

MH bulbs are much brighter and have a much larger coverage area than fluorescent light options; however, they require more power and run quite a bit warmer. MH bulbs are meant primarily for the vegetative stage and are between 4,000 and 6,000 Kelvin (K), resulting in a blue-colored light.

HPS BULBS AND DE HPS BULBS

For years, HPS lights—which give off a distinctive orange glow—have been an industry standard when plants are in bloom, as they can be adjusted from 400w to 1,000w depending on your ballast (with some ballasts allowing the bulb to put out over 1,000w). They can be cooled by air forced through the hoods they occupy. There are even water-cooled fixtures available for the HPS bulbs, but those have never really caught on.

Many different shapes and sizes of HPS bulbs are available that can work in magnetic and digital ballasts. The bulbs are around 2,200 K and can sit as low as 18 inches (45.75cm) above the canopy, depending on the hood they occupy and the other climate dynamics in the room. With the right configuration, these bulbs cover approximately a 4 × 4-foot (1.25×1.25m) footprint.

However, if you're looking for HPS bulbs that increase output by quite a substantial amount while requiring much less power and covering a larger footprint, you can try double-ended (DE) HPS bulbs. DE HPS lights include two bulbs in one reflector. Whereas older HPS lighting configurations would need to be lowered and raised according to your crops' canopy and rate of growth, DE HPS bulbs can remain stationary. However, you'll need at least 10-foot (3m) ceilings to maximize the potential of these bulbs. While some growers have utilized these lights in rooms with less than 10-foot (3m) ceilings by implementing certain low-height growing techniques (such as scrogging), the air conditioning still must be increased to compensate for the extra heat generated in the smaller grow space.

Hardening Your Plants with HPS Bulbs

Some growers utilize HPS lights in their grow room for the vegetative stage by adjusting the ballast to a lower wattage setting and leaving the timers in an 18- to 24-hour "on" period. This may not be the most efficient way to do this, but it works to harden your plants prior to transitioning into flower.

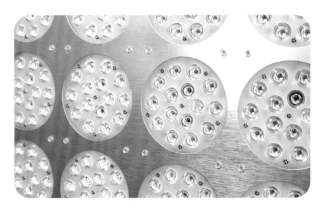

LED Bulbs

Advances in LED technology have been exponential during the last decade, resulting in a 10× decrease in cost and a 20× increase in lightwave efficiency. The quality of construction has improved, and the performance has continued to increase. LED lights last longer and provide more light than standard HID bulbs, using up to 30 percent less power.

Modern LED lights provide extremely high lighting uniformity and canopy penetration. This reduces the tendency for plants to waste energy "stretching" toward the light.

LED lights provide lower plant temperatures, too. Under an old-fashioned HID light, the temperature of a plant's leaves could be 10 to 20 degrees warmer than the surrounding ambient air. Under a modern LED light, the leaf temperature is only about 2 or 3 degrees warmer. This reduces much of the transpiration of moisture out of the plant, allowing it to retain moisture longer.

Today, LED light fixtures produce blended spectrums of light, which allow growers to select the "color," or wavelength, of light necessary for whatever stage of growth plants are in, while eliminating unnecessary "white" light.

For indoor growing in a limited space—especially in regards to energy usage, bulb efficiency, and temperature control—LED lighting, although it may cost a little more, is the way to go.

SOURCE	WATTS	DISTANCE ABOVE PLANTS
Fluorescent	T5	2 inches (5cm)
	T8	3 inches (7.5cm)
	T12	4 inches (10cm)
CFL	55 to 65	1 or 2 inches (2.5cm to 5cm)
	80 to 120	2 or 3 inches (5cm to 7.5cm)
	150 to 200	3 or 4 inches (7.5cm to 10cm)
LED	90	12 to 24 inches (30.5cm to 60cm)
HID	150	12 inches (30.5cm)
	175	12 inches (30.5cm)
	250	14 inches (35.5cm)
	400	18 inches (46cm)
	600	20 to 24 inches (50cm to 60cm)
	1000	24 to 36 inches (60cm to 90cm)
	1100	24 to 36 inches (60cm to 90cm)

This table lists how far away you should place your bulbs based on their type and wattage. As you can see, fluorescents can be situated quite close to your plants, while LED and HID bulbs noticeably need to be more distant. This is because fluorescents put out much less heat than HID bulbs and LEDs. They also don't have the output the others generate, so they need to be closer to be as effective as possible.

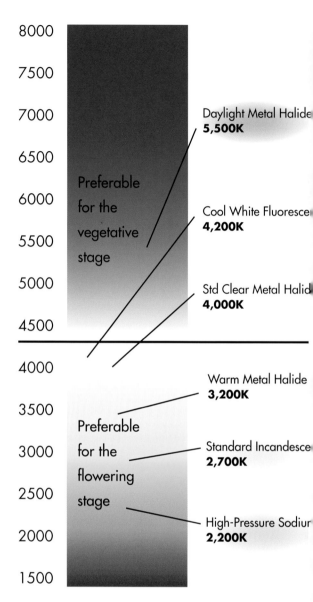

This chart shows the types of bulbs you should use for the vegetative and flowering stages. During the vegetative stage, you should use bulbs that give off blue and white light, such as fluorescent bulbs. This helps control the height of your plants and simulates the sun during spring and summer (the early stages of growth). During the flowering stage, you should use bulbs that give off yellow and orange light, such as HPS bulbs. This promotes bud growth and simulates the sun during fall and winter (the later stages of growth).

Hybrid Systems

Many growers, realizing the pros and cons of each type of lighting configuration, have implemented hybrid or supplemental lighting systems. Some grow structures use these supplemental lighting systems to increase productivity of their crops. Two popular types are HPS lights with fluorescent side lighting and LED-HPS hybrid systems.

Ballasts

Similar to how an amplifier works with speakers, ballasts power the bulbs of your lighting system. Their purpose is to provide the lighting system with high voltage during startup, and then to stabilize the arc by limiting the electrical current to the lighting system during operation. The more efficient the ballast, the more efficient your lighting configuration will be. There are basically two types of ballasts: magnetic and digital.

Magnetic Ballasts

Magnetic ballasts are an older technology comprised of one or more aluminum or copper coils that sit on a core made up of steel laminations. A transformer works in conjunction with a capacitor and sometimes an igniter. In terms of wattage, magnetic ballasts can switch between 100v and 220v. Unlike the newer digital ballasts, they tend to be heavier and run a bit hotter.

One of the advantages of having magnetic ballasts is their simpler components, which make them more durable and easier to repair. With a little know-how and the right parts, many growers are able to fix their magnetic ballasts, saving time, money, and crops. A downside to owning magnetic ballasts is the annoying buzzing sound they produce. But while the sound can be bothersome, it's basically harmless.

Digital Ballasts

Digital ballasts use microprocessors to control the electric current in your lighting system, which can be regulated through the use of firmware. This more-efficient setup allows them to be switched not only from 110v to 220v, but also to be adjusted to different wattage options. You can use digital ballasts to dial in the grow cycle, with some ballasts even having an option to implement a sunrise and sunset. You can also use the wattage selector to harden off your plants when going from the vegetative stage into the flowering stage. This versatility gives you much greater control over your crop than with magnetic ballasts.

However, digital ballasts tend to cost more. Plus, the number of components in digital ballasts makes them more likely to fail, and repairing them won't be as simple as repairing magnetic ones. Therefore, it's wise to keep backup ballasts on hand in the event of failure.

DE Bulb Ballasts

There are now ballasts that power double-ended (DE) bulbs. These new configurations illuminate the grow room like never before and save substantial amounts of power. No sound, much less heat, and brighter light make this new technology much more appealing.

Where to Keep Your Ballasts

Because a lot of heat is generated by the ballasts, you may prefer to keep them outside of your grow room. The downside to this is the lack of accessibility should you need to troubleshoot your lighting configuration; however, this can easily be mitigated by a simple labeling system. Simply label the light and corresponding ballast the same so you'll be able to navigate through the menagerie of wires and cords that most grow rooms seem to have.

Reflective **Hoods**

Now that you've learned about bulbs, ballasts, and lighting in general, you need to think about the way you'll focus or direct your light. If the light from your bulbs aren't focused, it won't be as concentrated as it could be, allowing valuable light to dissipate without ever being utilized by your plants. Imagine a lightshade and how it funnels the light into a certain direction, concentrating the light's output. You want to have that same effectiveness when it comes to lighting over your crop. Some growers actually dangle their light from the power cord without using any type of light-focusing device, but a few, more efficient options are available to you.

Closed Hoods

A closed hood reflects the light downward toward the canopy of the crop. This enhances the light's efforts, as the bulb is not only focused and directed, but also intensified by bouncing off the reflective material inside the hood. When using a closed hood, you should monitor the temperature below the hood and keep your plants at least 18 inches (45.75cm) away. That way, your plants won't be damaged due to excessive heat caused by the light. You can use a simple thermal gun to measure the temperature and adjust as needed.

Vented Hoods

Although the right amount of air conditioning will mitigate any heat put off by the lights, another option is a vented hood. A vented hood serves the same purpose as a closed hood, except you have the ability to push or pull cool air through it. This cools the bulb and allows the light to get closer to the crop's canopy. While some growers claim the photosynthetic active radiation (PAR) is reduced due to being pulled away by the air passing through a vented hood, testing has proven that to be a minimal variance at most.

Wings

Another reflective favorite of indoor growers is a wing. A simple wing-shape design (as the name implies) works well as a reflective hood, provided you have your room's temperature under control, as a wing can potentially cause hot spots on your plants. A wing is lightweight and inexpensive, and provides good distribution of light. You can also invest in a double-wing configuration, which greatly improves the coverage area.

Parabolic Reflectors

Similar to an umbrella in shape, a parabolic hood spreads the bulb's light over a larger area and works great for a garden utilizing supplemental lighting techniques. Because of the larger area of distributed light, it's best to overlap this type of reflective hood to ensure optimal penetration of light throughout the garden.

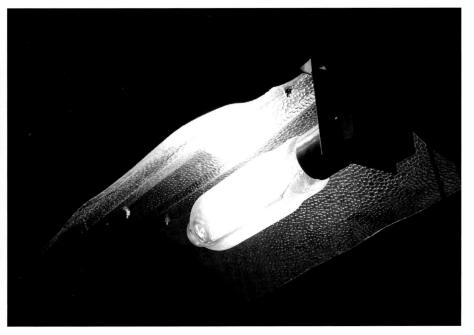

WINGS
The open design of wing reflectors blocks less light, allowing you to get the most out of your lighting.

Timers

"Timing is everything," they say, and when growing plants that flower in artificial conditions, it couldn't ring more true. The lighting systems for your indoor grow require timers to turn them on and off. This makes having good, functioning timers essential, as the difference between functioning and failed timers will be the difference between the success and failure of your garden.

Types of Timers

Because one simple failed timer could prove disastrous to your garden's success, selecting the right timer for your lighting applications is important. There are many types of timers and timing configurations, but the following are the three most common:

Mechanical timer. While you may think of the cheap 60-minute timer, the type you'll use for lighting is more sophisticated and has a 24-hour range. The operation is the same; you manually turn a dial to set it to the length of time you need the lights on, making it easy to use. However, because it works on clockwork and a wind-up system, you could have potential issues in precise timing and breakdowns in the structure.

Hardwired timer. This timer is wired directly to the electrical lines. It acts like a gatekeeper by connecting the circuit when the desired time is reached and disconnects the current when the timer goes off.

Digital timer. This timer has special electronics that can achieve a higher precision than mechanical timers. However, its cost is higher, with the potential for costly repairs if it stops working.

Timing the Cloning Stage

When cloning, the lights need to be on between 18 and 24 hours. I keep my cuttings under 24 hours of light to keep the cuttings warm and not disrupt the cycle in any way. However, anything more than 18 hours per day will suffice.

Timing the Vegetative Stage

As with cloning, your plants need to be under light between 18 and 24 hours per day in the vegetative stage. Although your plants will "sleep," or rest, regardless of whether the light is on or off, obviously they'd sleep better in the dark—and your electric bill will be less. That makes it ideal to operate under an 18-hours-on and 6-hours-off schedule.

Timing the Flowering Stage

Changing the lighting schedule is what truly initiates the flowering stage of plants. Once your plants have reached the desired size in the vegetative stage, you'll need to transfer them into your bloom area or simply switch the timers, provided you have sufficient lighting for vegetation and flowering. Although there are many schools of thought on what timing is best for your plants at this stage, replicating the strains' indigenous environment will produce the best results.

For beginners, a 12-hours-on and 12-hours-off schedule is perhaps the best until you feel confident in making adjustments throughout the flowering cycle. However, the room doesn't have to stay on the 12-hours-on and 12-hours-off schedule from the beginning of the flowering stage to the harvest. You can really manipulate the plant's output by dialing in the light cycle of your bloom room—setting your controllers to replicate sunrise and sunset, as well as shortening the daylight time at the end of the flowering cycle.

DIGITAL TIMER
These timers tend to have backup batteries, meaning they automatically reset in the event of a power failure.

Watering Methods

A plant takes in water and nutrients two ways: through the roots and through the leaves. While you can water only the roots or the leaves in several ways, watering in a way that lets you get both usually provides the best results. There are three methods for watering your cannabis plants: manual, automated, and foliar.

Manual Watering

Manual or hand watering allows you to deliver an exact amount of water to each plant. In soil, for instance, not all plants need the same amount of water during each feeding, and certain plants may not even need any water during the scheduled feeding time. Though not as scientific as an automated watering system, manual watering is the most natural method and allows you to interact with and observe your plants on a more frequent basis.

Automated Watering

An automated watering system includes a main water line, irrigation lines, water emitters, and a timer so all your plants are watered at about the same time. More practical than manual watering, automated watering is ideal for hydroponic systems. This is also a must when operating larger cultivation centers, as manual watering would simply take too long. The danger of automated watering is the crop usually shares water, which could potentially contaminate the entire crop if one plant becomes ill.

Foliar Watering

Spraying or foliar watering your plants allows them to take in nutrients through their leaves. Most foliar sprays promote node site growth or prevent pests, fungi, molds, and so on. This method is great for keeping cuttings fed prior to them developing a root system. However, foliar watering should be a supplemental nutrient delivery system to manual or automated watering, as your plants' primary mode of feeding is through their root systems.

Choosing a Watering Method

The method in which you water depends greatly on your growing method, the grow medium you use, and the environment in which you're growing your plants. For example, if you grow in soil, an automated watering method could potentially overwater or underwater some of your plants. However, using a manual hand-watering method for soil would allow you to individually inspect and assess each plant's watering needs.

Hydroponic Systems

If the soil in your area isn't conducive to growing marijuana or you have very little space in which to work, you can use a hydroponic system. There are six types of hydroponic setups: aeroponic, drip, DWC, ebb and flow, NFT, and wick. The following goes through how each of these systems work.

AEROPONIC SYSTEM

Aeroponic

This is a timed method that sprays water on the roots, which hang in the air, from the reservoir below by way of a pump. The water then returns to the reservoir, where it waits until the next watering cycle. Because this is a recycled watering system, it's wise to implement a water chiller to prevent the water from reaching dangerous temperatures over 80°F (27°C).

Drip

Like an aeroponic system, a drip system is a timed method relying on pumps. Highly effective and used in most major cultivation operations, this method feeds the plants through a drip stake, which drips nutrients onto the base of each plant. This forces the roots to seek water rather than the water seeking the roots. The drip system can be utilized as a recycled watering system, where the water returns to the reservoir after feeding, or a drain-to-waste system, where the water is disposed of after the watering cycle.

Deep Water Culture

Deep water culture (DWC) is another popular method of growing hydroponically. In this system, the water is pushed up through an air pump and the roots dangle in the water, allowing constant feeding. The risk with this method is water warming and root rot; however, this can be mitigated by connecting a water chiller and keeping the water below 70°F (21°C).

Ebb and Flow

Probably the most famous of growing techniques is the simple ebb-and-flow system. With this, water is pumped into the grow tray at timed intervals, temporarily flooding it with nutrients, and then drained after reaching the desired amount of time. A water chiller would be ideal with this system as well, as the temperature may get too high otherwise, leading to an increased need for oxygen from your plants.

Nutrient Film Technique

Nutrient film technique (NFT) is a method in which a constant flow of water is pumped into the grow tray from the reservoir, allowing dangling roots to feed. One of the most common and simple hydroponic growing systems it constantly feeds your plants, eliminating the risk of timer failure.

Wick

Definitely the simplest way to grow hydroponically, the wick method requires no pumps or timers. Instead, the plants simply pull what's needed from the reservoir by way of a wick. Similar to a wick in a candle, the wick is in both the water of the reservoir and the bottom of the feeding tray, connecting the plants to the beneficial nutrients.

DEEP WATER CULTURE

Setting Up a **Hydroponic System**

If you've decided to grow hydroponically, first ensure you have access to drainage and a water supply, as well as power for your light. You'll then need to select the system you'd like to use. As I've discussed, there are six types of hydroponic systems to choose from, and each has its own set of pros and cons. If you're a beginner in the world of hydroponics, it would be wise to select a system that will cause the least amount of frustration. Here, I discuss how to set up an ebb-and-flow table.

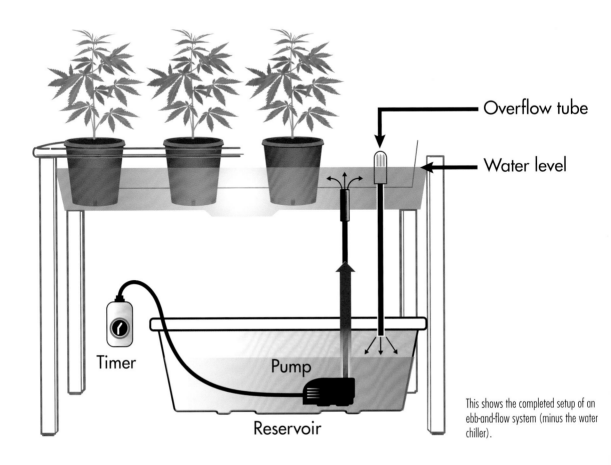

Overflow tube

Water level

Timer

Pump

Reservoir

This shows the completed setup of an ebb-and-flow system (minus the water chiller).

EQUIPMENT

4 × 2-foot (1.25×.5m) flood table
10-foot × 1-inch (3m×2.5cm) black
   hose
2 1-inch (2.5cm) drainage fittings
1-inch (2.5cm) PVC pipe about
   3 feet (1m) long
4-foot (1.25m) folding table
Air stone
15- to 20-gallon (56.75 to 75.75
   liter) hydroponic reservoir (larger
   is fine)
300-gallon-per-hour (GPH)
   submersible pump
10-foot (3m) airline hose
Air pump
Timer with 15-minute increments

TOOLS

Drill
PVC cutter or scissors that will cut
   the black hose

OPTIONAL EQUIPMENT AND TOOLS

The following items can help you set
up a water chiller for your ebb-and-
flow system, which is nice to have
but not necessary.

Water chiller ($1/10$ HP or more)
Hose for water chiller
260-GPH submersible pump
1-inch (2.5cm) hole saw
Power strip

What's an Air Stone?

An air stone is a piece of limewood or porous stone that
gradually diffuses oxygen into the hydroponic system in
small rather than large bubbles.

Step 1: Setting Up the Flood Table

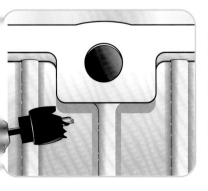

3 To create your overflow tube,
put the PVC pipe in one of the
drainage fittings.

1 Place the flood table face down.
Near the middle of each long
side, drill out 2 1-inch (2.5cm)
holes about 1 inch (2.5cm) from the
edge. (One hole will be for
drainage, while the other will be for
the pump.)

2 Cut the black hose in half and
attach one to each drainage
fitting. Attach the drainage fittings to
the flood table.

4 Put the flood table on top of
the folding table and allow the
drainage fittings to hang over the
edge.

Step 2: Setting Up the Hydroponic Reservoir

1 Put the air stone inside the hydroponic reservoir. Place the hydroponic reservoir on the floor under the folding table. Insert the free ends of the black hoses through the holes in the lid of the hydroponic reservoir.

2 Attach the 300-GPH submersible pump to one of the black hose lines. (Fittings should have been included with the pump that connect the hose to the pump.)

Step 3: Setting Up the Air Pump

1 Cut a length of airline hose long enough to go from the plugged-in air pump to the air stone in the hydroponic reservoir.

2 Attach the airline hose to the air pump. Attach the air stone to the other end of the airline hose.

3 Adjust the air pump so bubbles are coming to the top, but not too many.

Step 4: Setting Up the Timer

1 Set the timer for 15 minutes on and 45 minutes off every hour. (This will change as your plants grow older and develop longer roots.)

2 Plug the feed pump (the larger pump connected to black hose lines of the flood table) into the timer.

3 Fill the hydroponic reservoir with water. Turn the dial until the feed pump turns on. Once turned on, make sure the feed pump has no leaks.

4 Wait for 15 minutes to pass to ensure the table doesn't fill faster than it drains.

You now have a functioning constant ebb-and-flow table! While topping off—or adding water to the hydroponic reservoir—is fine, for the best results, change out the water in your ebb-and-flow table once a week.

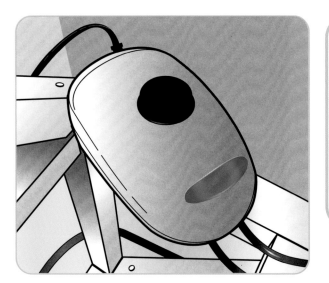

Nutrient Regimens

Once you've set up your ebb-and-flow table, you can find a good nutrient regimen and start growing. Follow the nutrient regimen throughout the vegetative and flowering stages, keeping in mind your water should be changed out each time your regimen changes.

Optional: Setting Up the Water Chiller

You can better control your water with the implementation of a water chiller. The following takes you through how to set up this external unit for your ebb-and-flow system.

1 Connect the hose for the water chiller to both of the water chiller fittings.

2 Connect the "in" hose to the 260-GPH submersible pump in order to pump water through the water chiller.

3 Put the other hose into the hydroponic reservoir. Use your drill and hole saw if additional holes are needed in the lid of the hydroponic reservoir for the in and out hoses.

Ventilation: **Airflow**

Ventilation systems have a direct effect on the function of your plants. Airflow is the total exchange of stale air in a room for fresh air. Good airflow helps regulate temperature and prevent mold and other plant issues. So a proper indoor ventilation system is the difference between success and potential failure of your crop.

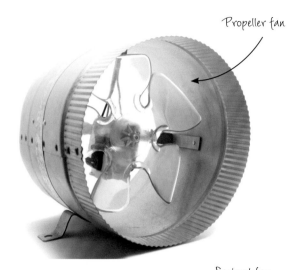

Propeller fan

Squirrel fan

Size and Season

Before figuring out the best airflow in your grow room, you'll need to make a couple decisions.

What size indoor space will I be growing out of? Smaller indoor grow rooms require equal airflow but will be easier to configure than larger grow rooms.

Will I be growing year-round? This is important because the temperature and humidity will change throughout the seasons. This will make it necessary to have airflow configurations for each grow season.

Calculating the Volume of Your Grow Area

When thinking about the airflow in your grow room, the first thing you need to do is figure out the volume of your space. To do this, simply multiply the length by the width (otherwise known as the square footage [square meters] of a space), and then multiply the square footage (square meters) by the height of the indoor space. If your indoor grow room has a pitched roof, you can go with a height of 10 feet (3m) for ease of calculation. This will give you a cubic footage (cubic meters) for your space. Because all fans are rated by the amount of cubic feet of air they move per minute (CFM), knowing the cubic feet of grow space tells you what size of and how many fans to buy.

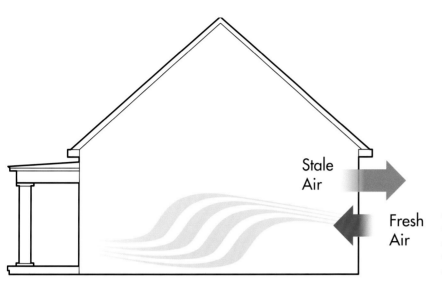

Stale
Air

Fresh
Air

Airflow in your grow room is all about exchanging air—taking out stale air and bringing in fresh air. As this image shows, one of the reasons you may do this is to cool down your indoor grow room, exchanging warm air for cool air.

Airflow by Season

Depending on how well your indoor grow room or grow tent is insulated from seasonal variables, different seasons require different or more powerful equipment. A consistent climate is necessary for a consistent harvest of quality cannabis.

FOR SUMMER

During the summer, more fans are needed to move air than in the other seasons. Dehumidifiers or humidifiers might also need to be implemented; however, this depends greatly on the climate outdoors. One air exchange per minute is the general rule of thumb for the minimum rate of ventilation during the summer months. Most ventilation efforts are focused on keeping an indoor grow room cool, while a proper air circulation plan (discussed next) will help lower humidity.

FOR SPRING AND FALL

The ventilation rates during the spring and fall will be somewhere in between the summer and winter rates. It probably won't be as hot or humid as summer, and it more than likely won't be as cold as winter. You can get away with fewer fans and possibly a heater to maintain the best airflow. If you prepare your indoor grow room for the summer months, you should be fine for the rest of the year, as those are usually the most difficult months to control. An automated ventilation system that's controlled by thermostats is most efficient during these months.

FOR WINTER

At this time of year, an air exchange should occur three times per hour or once every 20 minutes. During the winter months, ventilation efforts transition from temperature control to humidity control due to the fact that the temperature difference from the inside of your grow room to the outside creates condensation. The humidity created because of the condensation can be catastrophic if it's not removed. Plus, most indoor grow rooms require supplemental heat during the winter months, which further complicates the humidity issues. To remove the humidity, you'll need to calculate a rate of ventilation that won't defeat your heating efforts.

Ventilation: **Air Circulation**

One of the most misunderstood and underimplemented steps in growing is proper air circulation, or the movement of air within a space. Plants need fresh air or CO_2 just as people need fresh oxygen to stay alive. If you're not planning on providing fresh CO_2, you'll need to exchange the air in the room according to a formula derived from the size and volume of the room. On average, the total volume of air in a grow room or tent should be exchanged at a rate of around every 5 minutes.

Indoor Ventilation

In an indoor grow that isn't sealed well enough to implement CO_2, you'll need to exchange the air on a timed schedule. This can be done by way of a timer that's connected to a fan and air damper. The fan and damper are connected to a carbon filter and all three are inside the room. This configuration will pull air out of the room through the carbon filter, while a separate fan pulls filtered fresh air in through elsewhere in the room.

Carbon filters are necessary only if you want to eliminate the odor of cannabis from the air being pulled out of your grow room. Fresh air that's being pulled into the room should be filtered to prevent pollen, insects, and other contaminates from entering.

OSCILLATING FAN
Wall-mounted oscillating fans are a great way to circulate the air above the canopy of your grow. This also enables your plants' limbs to grow stronger in preparation for bearing buds.

Closed Grow Rooms

Because a closed grow room is completely airtight, the CO_2 levels are easier to maintain and control. You can make this happen by simply pumping in the CO_2 with tanks and mixing it with fans. As long as your AC isn't exchanging air outside the room, you'll then be good to go. The best ACs to use in this scenario are mini splits because their ductless design provides a more targeted movement of air. This makes the air circulation process more efficient.

Grow Structure Ventilation

Grow structure ventilation is a little different, and the type of grow structure you're cultivating in will determine what kind of air circulation is best. There are basically two types of ventilation: mechanical (active) and natural (passive). A natural ventilation system has no powered equipment but rather operates on the concept of thermal buoyancy. Hot air rises, so you can use this scientific principle in your favor. A series of sidewall vents and roof vents will cause thermal buoyancy. When it gets hotter in the grow structure, the hot air rises up and out through the roof vents while pulling in cooler air through the sidewall vents. Positioning your grow structure to manipulate the natural prevailing wind will allow you to utilize this technique most optimally.

Supplying
Carbon Dioxide

Carbon dioxide (CO_2) is a colorless and odorless gas used by plants to create energy from light. Employed properly, adding CO_2 can potentially bolster your plants' growth.

Advantages and Disadvantages of Using CO_2

Before we get into in what ways you can employ CO_2 and how, let's take a look at the advantages and disadvantages of using CO_2 in your grow room. Obviously, I think the benefits outweigh the costs, but this will help you decide for yourself whether supplying CO_2 is right for your grow room.

ADVANTAGES

You get better plant growth and yield. In a bright indoor grow environment, there's typically more light than your plants can naturally use. Supplying CO_2 can help them use more of your grow room's light, which can cause your plants to grow up to 50 percent faster and your overall yield to increase by 10 to 20 percent.

You can choose from a variety of delivery systems. CO_2 can be delivered to your plants in many different ways, from dry ice to tanks. This gives you the benefit of choosing a system that works best for your grow space at a cost you can afford.

You only have to use it during light cycles. Because CO_2 aids plants in creating energy from light, your plants won't need CO_2 at night. This makes it easy to schedule when you're using CO_2 and determine how much you'll need for your grow room.

DISADVANTAGES

Your grow room has to be well sealed. In order for CO_2 to produce the results you're looking for, your grow room has to be sealed. Otherwise, CO_2 will leak out of the room, which defeats the purpose of pumping it in and increases costs to resupply it.

Your plants become more high maintenance. While the payoff of a bigger crop of larger plants is ideal, your plants will in turn need more of everything—nutrients, water, and possibly even space—in order to stay healthy.

More isn't necessary better. There is such a thing as giving your plants too much CO_2. This can lead to the opposite effect of what you're hoping—a slowdown in growth and even plant death.

CO$_2$ Delivery Systems

In order for CO$_2$ to be effective, you need to maintain levels of 1,200 to 1,500 parts per million (ppm) in your indoor grow room. You can use either a passive (no mechanical assistance) or active (mechanical assistance) delivery system for CO$_2$. Different delivery systems provide CO$_2$ at different rates, so be sure you know what those rates are before choosing your system.

PASSIVE DELIVERY SYSTEMS

Seltzer water. Seltzer water is an inexpensive method that emits CO$_2$ through carbonation. It's typically sprayed directly on the plant leaves (while being careful not to directly spray the buds) to help facilitate growth. However, it's inefficent, as the water can raise humidity and will evaporate quickly under the lights, preventing plants from absorbing as much CO$_2$ as they could.

Dry ice. Dry ice is a relatively inexpensive method for delivery of CO$_2$. Hanging dry ice in a bucket above the plant allows the dry ice to shower the plant in CO$_2$ as it evaporates. Typically, your personal-use marijuana growing space will use about 1 pound (450g) of dry ice a day, give or take, depending on environmental conditions. However, because it doesn't last long, dry ice is only effective for short-term use.

CO$_2$ bags. CO$_2$ bags are a self-contained system for delivery of CO$_2$ through the process of fermentation, in which fungi grow on organic matter. CO$_2$ bags typically last several months and can be purchased at most garden stores. However, this method requires many bags, as well as a sealed room, to be efficient.

CO$_2$ BAG
This uses fungi that grows on organic matter to produce CO$_2$.

ACTIVE DELIVERY SYSTEMS

CO_2 tanks. CO_2 tanks are an efficient way to enrich grow spaces. The equipment for a pressurized CO_2 tank delivery system includes a cylinder tank of compressed CO_2 gas, a pressure regulator to maintain consistent pressure, a flow meter to regulate the amount of gas being released, an automatic on/off valve, a variable timer to ensure that gas is emitted during intervals that will maintain CO_2 levels during the desired time (synced, for instance, with the closing of ventilation), and additional assorted connectors (hoses, adapters, and so on). While some high upfront cost is involved for the necessary equipment (typically around $500), you'll have great control over the level of CO_2 in your grow room.

CO_2 generators. CO_2 generators burn a fuel such as propane, butane, or natural gas, which produces CO_2 and water vapor. Commercial-size CO_2 generators are too large and expensive for personal marijuana growing spaces. However, smaller devices—such as Bunsen burners, small gas stoves, fuel-burning camping lanterns, and propane-powered heaters—can all be used to generate CO_2 as the fuel is burned. One pound (450g) of fuel will produce about 3 pounds (1.5kg) of CO_2. A downside of burning fuel for CO_2 via a CO_2 generator is it will also produce more than 20,000 British thermal units (BTUs) of heat, which is generally not necessary or recommended for climate-controlled grow rooms.

Conditions and Timing for CO_2 Application

To be utilized properly, environmental conditions for plant growth must be controlled at an ideal temperature of 75°F to 85°F (24°C to 29°C) and an ideal humidity of about 40 percent. Additionally, during the application of CO_2, all fresh air intake and exhaust fans should be closed or shut off respectively. This keeps the extra CO_2 you've applied inside the grow room and near your plants.

Plants use CO_2 in the process of photosynthesis, helping to convert light into energy for the plant. Therefore, it is only during the period that your grow lights are on that CO_2 needs to be used, as plants don't undergo photosynthesis in the dark. Depending on whether your delivery systems are active or passive, timers can be employed to deliver CO_2 during those times when the lights are on, the room is sealed, and the intake/exhaust fans are off.

Danger Zones for Your Plants' PPM

Cannabis plants can easily absorb up to 1,500 ppm of CO_2. Without enough CO_2 (less than 200 ppm), plants can't photosynthesize light. With too much CO_2, (more than 2,000 ppm), CO_2 becomes toxic to marijuana.

PASSIVE CO_2 DISPERSAL CANISTER

Simply add water to this all-natural passive delivery system to emit CO_2 for approximately 2 weeks.

Growth Cycles

Staggering growth cycles of several plants is an effective way to ensure a monthly harvest of usable marijuana for personal use all year round. Typically, this is done indoors using a constant rotation of plants grown from clones or seeds, plants in the vegetative stage, and plants in the flowering stage.

Timing Harvests

You can determine the timing of a regular rotating harvest using an average of 4 weeks for cloning or sprouting seeds, 4 to 6 weeks in the vegetative stage, and 8 weeks in the flowering stage.

- Rooting clones or seedlings: 4 weeks

- One younger, immature plant in the vegetative stage: 4 weeks from the flowering stage

- One older, mature plant in the vegetative stage: ready for the flowering stage

- One plant in the first 4 weeks of the flowering stage: 4 weeks until harvest

- One plant in the last 4 weeks of the flowering stage: ready for harvest

Space and Time Considerations

When deciding whether this method will work for you, consider the logistics of space and time in regard to rotating growth cycles. It will be necessary for you to have a space for both the vegetative growth cycle and the flowering cycle operating simultaneously. Also, maintaining several stages of plant growth at once—from seedlings to mature flowering plants—will demand more of your personal attention and time.

Rotating Growth Cycles

The typical growth cycle of a cannabis plant is at least 3 months from a rooted plant to the harvesting of buds. Assuming you're growing 4 plants, that means you could harvest four times a year more or less, for a total of 16 plants.

By taking clones from mature vegetative plants just prior to them being moved into the flowering stage, you can be sure a steady stream of plants will be reintroduced to the vegetative stage and ready for the flowering stage.

4 TO 6 WEEKS INTO FLOWER

6 WEEKS INTO VEGETATION

2 WEEKS INTO VEGETATION

Cleaning Indoors

The goal of cleaning is to keep your plants healthy. However, because molds and fungi can also be detrimental to property as well as personal health, it's of primary importance to keep your grow area clean and free of potentially harmful conditions, especially if you're growing in your home. Cleaning tools, removing dead plant matter, eliminating standing water, and getting rid of dirt and dust to ensure constant, fresh airflow are prudent and necessary measures for maintaining healthy plants.

What Should You Clean?

Everything used in the process of growing marijuana can and should be kept clean. This includes gardening and cutting tools, work areas, doorknobs, glass, ducting, reflective hoods, lightbulbs, and so on. All surfaces of your growing areas, including walls and floors, should also be cleaned as necessary.

Cleaning Your Equipment

The following are some things you can do to keep your grow room equipment clean:

- Soak your cutting tools in solutions of alcohol and water when still being used for cutting, or in bleach and water when not in use.

- Wash garden tools after use to remove dirt and prevent the potential spread of disease.

- Wash and wipe clean any surface areas you use when working with your plants—such as tabletops, doorknobs, and floors—with an alcohol solution.

- Use a duster or damp towel to keep ducting and reflective hoods free of dust.

- Use glass cleaner and paper towel to clean lightbulbs. (However, make sure the power is off and the temperature of the bulbs is cool to the touch before doing so.)

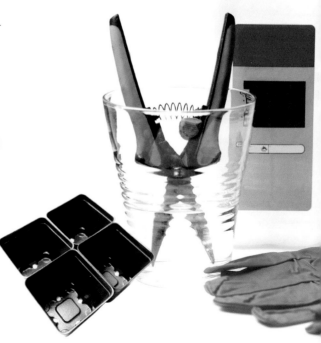

DEAD LEAF REMOVAL
Remove any dead leaves you find, either in pots or on the floor. They could harbor insects or encourage the growth of mold.

Alcohol Solutions

Because of its high alcohol content, inexpensive vodka makes a good cleaner. You can mix together a solution of 50 percent vodka and 50 percent water and use this for most of your disinfecting of surfaces and work areas. For mold and mildew remediation, simply use a straight solution of vodka and let it soak for several minutes before removal. Whatever solution you're using, always be sure to remove your plants from the area being cleaned. Cannabis plants are sensitive to alcohol- and bleach-based solutions.

Part 4
Outdoor Growing

For those with the space, outdoor growing mitigates
many costly issues associated with indoor growing.
However, there are still issues that must be considered
before you decide to grow outdoors, such as the
unforgiving nature of the elements. This part takes you
through the process of outdoor growing and details
how to deal with certain problems that come up in
these particular conditions.

Introduction to **Outdoor Growing**

Outdoor growing allows cannabis plants to thrive in an all-natural environment, where plants are naturally designed to grow. The natural elements of earth, wind, water, and sun provide benefits that can be difficult to replicate indoors. However, nature can sometimes be uncooperative in terms of the ideal growing conditions.

Pros

Growing outdoors is a sustainable contribution to a living ecosystem. Good soil is capable of sustaining all of the nutrients necessary to produce some of the largest root systems and best-tasting buds found on Earth. The size and health of a plant's root ball are directly related to the size and health of the flowering buds, providing a foundation for support and strength as the plant grows up and out while also providing a healthy supply of nutrients and water.

The sun is the most powerful light source available for your plants. The sun delivers a full spectrum of light efficiently and equally to all the flowers of a plant, unlike lightbulbs, which lose strength based on the distance from the plant. In some cases, producing great marijuana outdoors can even be as easy as putting a plant in the ground and watching it grow—if the conditions are right.

It's easy to make the conditions work for outdoor growing. In cases where the soil isn't ideal for gardening, fertile mixes can be added to supplement the dirt or a hole can be dug and a premixed growing medium used to fill it. Plants can also be potted and placed outside above the ground, allowing them to enjoy the free airflow and sunlight provided in outdoor conditions.

Cons

You can't control the weather. One of the most common concerns for an outdoor grower is the uncontrollable aspect of weather. Heavy rains, excessive winds, and extremely hot summer days can affect the growth and overall health of a plant. However, a few simple precautions—like a tarp for water, a stake for support, or a shade cloth—can mitigate most of the problems you may encounter.

It could be illegal to grow outdoors. Additionally, not all states that allow home growing for adult personal use specifically allow for marijuana plants to be grown outdoors in view of the public in general or even in view of your neighbors. If that's the case, you'll have to grow your marijuana indoors.

It can require a lot of space. Growing plants outdoors can sometimes require a lot of space. Some plants can grow up to 8 feet (2.5m) tall and 8 feet (2.5m) around.

Climate and Cannabis

When growing outdoors, climate plays a huge role both in what you contribute to the environment and what you plan to grow.

Environmental Impact

Unlike outdoor growing, indoor cultivation actually contributes to poor climate conditions by massive power consumption and water usage. The carbon footprint created by indoor grow operations definitely isn't helping the environment. In addition, many of the nutrients used in most indoor grows are actually considered hazardous waste once they've been mixed together. These are harmful to the environment, and more times than not, they aren't disposed of properly.

With outdoor growth, however, you have a chance to contribute positively to the environment. Trees and plants are a much-needed resource here on planet Earth, and with constant deforestation occurring around the globe, the more cannabis plants growing outdoors, the better. Given the fact that the air we breathe is comprised of only 19 percent oxygen, cannabis is a great option to offset the damage caused by pollution, as it's a hearty crop that grows like a weed and replenishes quickly after harvest.

Cultivating in Warmer vs. Colder Climates

Just as people from various parts of the world have different physical characteristics, so do cannabis strains. Varieties found in dry, mountainous regions are very different than those indigenous to tropical climates. So based on your climate, you must consider which strains will wilt versus thrive.

It's much easier to cultivate in warmer climates, so you'll have a much larger choice of genetics to choose from when deciding which strain to grow in your garden. However, the best types are sativa varieties, as water doesn't evaporate from their leaves. This allows sativas to grow healthy even in the driest, hottest conditions. AK-47, Amnesia Haze, and White Widow are just a few strains that grow well in hot climates.

If you're growing in colder climates, you won't be able to use auto-flowering plants. Instead, you'll need seeds with modified genetics in order to grow the best plants in the shortest amount of time. Indica varieties tend to be best for cold weather, as their compact nature keeps cold air from entering. Super Skunk and OG Kush are examples of strains that endure well in colder climates.

Growing Cannabis at Disaster Sites

In Chernobyl, where a catastrophic nuclear disaster occurred, hemp fields have been planted to absorb the leaked radiation. This is also being discussed as a possible solution to contain some of the damage caused by the Fukushima nuclear disaster.

Choosing a Spot

Your outdoor marijuana grow will benefit the most from a location that offers plants the most sunlight possible, has a good soil medium, is protected from animals, and remains out of the view of the general public.

Planting in Containers vs. the Ground

In the ground

In a container

When deciding to plant your marijuana, whether in a container or in the ground, the most important thing to consider is the soil medium. Nutrient and water uptake, drainage, and root foundations are all influenced by soil. Supplements—whether store bought or provided by growing a season of cover crops that decompose naturally— can help enrich the soil for your plants.

If you plan to plant in the ground, planting clover or winter peas prior to a season of growing marijuana will provide nitrogen to the soil as they decompose, while planting buckwheat will provide phosphorus.

Containers can be used when ground soil isn't good for growing. However, the size and shape of the container limits the size and shape of the plants. As the roots grow, so grow the "fruit" (a.k.a., the flowers or bud). For healthy roots in containers outdoors, make sure the container is large enough to support extensive root growth and has adequate drainage to prevent overwatering.

Terrain and Terroir

Ideally, a southern-facing terrain with a slightly downward-facing slope allows your plants the most exposure to direct sunlight. Environmental conditions and other specific conditions that affect a plant's habitat are known as *terroir*. Ideal terroir includes soil rich in organic matter, a constant gentle breeze, and a natural humidity of about 40 percent.

Getting Some Sun

Sunlight offers the most energy possible for the growth of your plants. The strength of the sun's light isn't diminished between the top and bottom flowers of outdoor marijuana plants, meaning you'll get very even coverage if your plants are in direct sunlight. So avoid planting in the shade, such as under a tree or too close to other plants.

Putting Up Fences

Animals and humans alike will migrate toward the smell of a fresh cannabis plant. To protect your marijuana plants from cats, deer, dogs, and other prying eyes, fences are essential. Depending on the laws in your location—as well as who or what you're keeping out—you can have a fence as high as 8 feet (2.5m) tall.

Keeping Plants Out of View

Position your marijuana plants out of the view of the general public to prevent complaints or violations of law. This is easily accomplished by strategic placement on private property (for example, behind a garage) or sight-obscuring fencing. Please check your local laws to determine your rights to grow outdoors.

A tall fence can help keep out predators.

Choosing Strains for Growing Outdoors

When deciding which strains to grow outdoors, you have several factors to keep in mind. While evaluating your options, it's best to learn about where you'll be cultivating and find strains that are indigenous to similar areas.

Climate

As I've already discussed, climate is a huge factor when choosing the cannabis strain you want to grow. The lack of control you have over the outdoor climate means you'll have a different set of deciding factors when selecting a strain to cultivate versus for indoor growth. In general, you'll want heartier, larger varieties that are less prone to disease and pests. While sativas and hybrids are usually the easiest to cultivate outdoors, a hardiness map can help determine which strains match best with the climate you intend to grow in.

Temperature

The temperature of your grow area should be similar to the region of the world in which your intended strain comes from. If you live in a colder area, you'll be limited as to which strains will survive in that type of environment. However, if you're going to grow in a warmer climate, your options increase dramatically. For instance, indicas thrive in cooler climates, while sativas thrive in warmer climates. Whatever you choose, make sure the temperature you intend to grow in doesn't fluctuate too much from day to night so you'll have a relatively stable environment for your plants.

Humidity

Another important consideration when deciding which strains to grow outdoors is the relative humidity (RH). If you're going to grow in a humid environment, you'll probably want to steer clear of plants that have Kush lineage, as they are much more prone to powdery mildew. However, those same types of plants, along with most indicas, do well in less humid environments.

Shorter Growing Seasons

If your season isn't long enough for a 10-, 11-, or 12-week flowering stage— typical of colder climates—you'll be better off with auto-flowering indica varieties, which can be harvested in as little as 8 weeks.

Popular Strains for Growing Outdoors

Even after narrowing down your choices by climate, temperature, and humidity, you'll still have hundreds of strains to choose from. If you're unsure where to start, the following are some popular strains cultivated by marijuana growers outdoors.

Afghanica. This hybrid strain is hardy and does well in colder climates. It has a relaxing high and a sweet, skunky taste.

Big Blue Cheese. A potent and high-yielding indica-dominant strain, it can be planted outdoors in colder climates. This strain has a complex flavor of berries, spice, and cheese.

Blueberry. This indica-dominant hybrid grows strong and sturdy, making it capable of standing up to harsh weather conditions. It has a berry flavor and aroma.

Green Crack. This sativa grows well in a warm, sunny climate. It has a euphoric high with a spicy and citrusy flavor.

Mango. This sativa-dominant strain has a pleasing mango taste with a sweet smell. It grows particularly well in warmer climates.

Maui Wowie. This sativa strain grows well in tropical climates and is pretty resistant to mold and diseases. It has an energetic high and a sweet pineapple taste and smell.

Northern Lights. With an auto-flower time of 6 to 8 weeks, this indica-dominant strain is hearty enough to grow in almost any climate. Providing heavy body highs that motivate you to lay on the couch, it's a good choice if you want to relax.

Available Sunlight

The position of the sun will move throughout the seasons, so be sure sunlight remains available throughout the intended flowering stage of your plants. A slope with a southern exposure is ideal for outdoor growing during this time.

Soil Preparation

Preparing your soil properly will ensure a healthy foundation for the growth of your marijuana plants. Just as with your plants, a healthy soil medium should be nurtured with care and attention.

Tilling

Tilling your cannabis garden can be done easily with a rototiller or garden hoe. Tilling loosens the soil to allow easier digging and, more importantly, facilitates drainage and growth for your plants' roots. Tilling also allows your soil medium to be amended with additional nutrients or aggregate, if needed.

Checking the pH

The balance between acidity and alkalinity is determined by checking the pH. Marijuana prefers a slightly acidic pH of between 6.5 and 6.8 when grown in a soil medium, allowing efficient uptake of nutrients by the plant. The water and dirt can be tested independently; however, the most accurate measure of pH is after the dirt and water have been mixed together as they would be when watering a plant under normal conditions. Electronic pH meters are inexpensive and provide accurate measurements of the pH instantly.

Soil Testing

Determining the quality of your soil is accomplished by testing. Testing is inexpensive and can save you money by efficiently informing you of exactly how much of which nutrients and fertilizers your soil needs for optimal plant growth. Soil testing kits can be purchased at most plant nurseries and big-box hardware stores. You can also have the testing performed by your local conservation district.

USING AN ELECTRONIC pH METER
Insert the meter into the soil and keep it in there until the time listed by the manufacturer (typically about 1 minute).

Digging Holes

When digging holes to plant, they should be large enough to accommodate both the size of a plant's root system plus any additional amendments your soil may require.

Dig holes using a spade, shovel, or backhoe. Small holes can be dug with a hand spade to accommodate seedlings and small plants; for larger plants, however, use a shovel or a backhoe. When determining what size hole to dig, consider the size of your plant at the time of planting and the length of your growing season. For instance, a hole big enough for a 1-gallon (3.75-liter) pot, if planted early spring, will allow a plant to grow throughout the entire summer into fall and could be quite large.

Amend the soil either by mixing in nutrients prior to backfilling or by adding amendments directly to the hole. If your dirt and terrain isn't plant friendly, mix a potting soil medium to refill the entire hole. If your soil is healthy, simply mix the amendments with your existing soil.

Adding Water

When planting in dry soil, fill the hole with water prior to planting to soften the soil and activate any nutrients.

MEASURING THE HOLE
You can use a stick as a measuring device to gauge the depth of a hole.

Amending the Soil

Cannabis plants thrive in conditions very similar to tomato plants outdoors. Like with tomato gardening, soil mediums are often not ideal for marijuana gardening and, therefore, need amending. Nitrogen, potassium, potash, and phosphorus are common ingredients in healthy soils.

Peat/coco

Bat guano

Peat/Coco

Peat and coco are common soil amendments. Typically, one or the other is used exclusively as a base. Both offer good drainage of excess water while maintaining moisture for the plant roots. Peat or coco should be mixed (or purchased premixed) with at least 50 percent regular soil.

Aggregate

To facilitate water drainage, an aggregate of perlite (a volcanic rock) or pea gravel can be used, if needed. While up to 25 percent of your soil amendment can be mixed with aggregate, a simpler method is to place about 1 inch (2.5cm) of aggregate in the bottom of the hole or container.

Bat Guano

Bat guano provides the all-natural, beneficial nutrients and microbes needed during both the growth and flowering stages of your marijuana plants. Carnivorous and insect-eating bats produce guano with healthy amounts of nitrogen (which is great for plant growth), while fruit-eating bats produce guano higher in phosphorus (which is great for flowering). Add bat guano as needed to produce a healthy soil medium.

Liquefied seaweed

Mixing It Up

Soil mediums should drain easily, with an acceptable pH after adding water, and test well for both nitrogen and phosphorus. If they don't, the following amendments are needed.

1 Amend the soil with peat or coco in a 50/50 ratio by volume to facilitate water drainage, as well as provide an adequate medium for healthy root growth. This simple amendment will not significantly affect the pH balance or nutrient makeup of your soil in general.

2 If necessary, aggregates like perlite and pea gravel can be used in soil to facilitate water drainage. When adding to a soil or soil-peat mix, use a 1-to-4 ratio in your soil medium. If you're combining the aggregates with peat or coco only, use a 50/50 ratio by volume.

3 You can then add nutrients as necessary while monitoring the pH of your soil medium after watering.

Seaweed

Liquefied seaweed and kelp extracts have increased chlorophyll production, which facilitates photosynthesis. These extracts also encourage growth of beneficial bacteria, leading to better nutrient uptake. This mild supplement can be added liberally. Simply make sure your pH reading after watering reads between 6.2 and 7.0.

Hardening
Plants

You can begin and end the growing process completely outdoors, but you may choose to grow your plants indoors and then move them outdoors for flowering and beyond.

Hardening

Plants can't simply be moved from indoors to outdoors, though; they must be hardened. Hardening, or *hardening off*, is the process of exposing your plants to a new environment in increasing intervals to prepare them for the full transition to that environment. If you don't harden your plants, they could get stressed by the change, which will impede their growth. This technique is technically beneficial for both indoor and outdoor growers, but it was originally developed by growers who started growing marijuana indoors and moved their plants outdoors to bloom.

Why Make This Transition?

It may seem strange to talk about indoor growing in the part of the book about growing outdoors, but your plants can benefit from starting indoors first. One benefit of making this transition is being able to harvest more than once a year. When you're growing entirely outdoors, you have to plan so your plants' life cycle follows the seasons. However, beginning the growing process indoors allows you to grow plants without waiting for warmer temperatures and more hours of sunlight. Plus, by keeping them indoors during their early stages, your plants begin life in the safest, most controlled environment, free from colder temperatures, pests, and so on.

How to Harden Your Plants

In order to acclimate your plants to an outdoor environment, start by taking your vegetative plants out for a few hours a day. This will allow them to slowly get used to the climate they'll be experiencing on a full-time basis. Find a shady spot for the first few days so they aren't immediately exposed to harsh sunlight, and then slowly increase the amount of time outside until they're ready for the full transition.

While you're getting them used to being outdoors, you can't discount the conditions indoors. Your indoor environment needs to be properly set up so you avoid potentially stressing your plants on their return to it during this acclimation period. More closely matching the outdoor environment—for instance, with lighting that isn't too weak compared to what's outside—will help ease your plants' transition. Eventually, your plants can then move completely outdoors without any negative effects.

Watch Out for Windburn

When growing outdoors, plants are subject to windburn. Don't underestimate the damage that can occur from the wind outside on plants that haven't been exposed to heavier air movement. You'll learn more about how to protect your plants from this and other elements in the next section.

Protecting Plants from the Elements

Growing outdoors means you're sometimes faced with natural elements beyond your control. When growing outdoors, plants in containers and plants in the ground can be affected differently by the same environmental factors. Always listen to the weather report, and check out the following conditions for tips on what you can do to protect your plants.

Excessive Sun

Direct sunlight provides the best possible light for growing marijuana. However, long periods of exposure to direct sunlight with air temperatures in the high 90s (30s Celsius) and above can lead to heat stress. Heat stress causes leaves and flowers to curl upward and become deformed, stunting growth. A shade cloth, available in most garden stores, will prevent heat stress during extend periods of sunlight for air temperatures approaching 100°F (37°C). If you're growing outdoors in containers, keep the roots cool by placing the containers in the shade.

Wind

High wind can blow over plants in containers, possibly damaging branches. If possible, simply bring the containers inside. Plants in the ground can be secured to stakes to avoid damage. If high wind is a constant environmental factor, a windbreak should be utilized to avoid stress-related damage. Planting a hedge of evergreen arborvitae offers a good wind break, as do solid fences or walls. However, don't let your windbreak shade your plants.

Excessive Rain

Rain is the most efficient way for your plants to get water; however, it can also cause various issues with them. Soil can wash away, so be mindful of how surface water flows in your outdoor garden. Also, to avoid saturation from water, ensure there's proper drainage around your roots. Another concern is small plants, which can be broken by heavy rains and/or hail. If they're grown in containers, they can be brought inside during brief periods of heavy rain.

Excessive moisture will cause mold to grow in the buds during the flowering phase of cannabis plants. In brief periods of heavy rain, a garbage bag will keep maturing buds dry. However, if extended periods of rain are forecast, a clear plastic tarp or hoop house will keep plants dry and allow adequate airflow.

Cold

In early spring and late fall, when temperatures approach freezing at night, plants should be covered with a cloth or garbage bag to prevent damage from cold and frost.

EXCESSIVE SUN CAN CAUSE PLANTS TO CURL UPWARD AND DRY OUT.

OP HOUSES CAN KEEP YOUR PLANTS DRY DURING RAIN.

SECURING YOUR PLANTS TO STAKES HELPS AVOID WIND DAMAGE.

Watering Outdoors

When cultivating outdoors, your watering schedule will be dependent upon the rainfall where you're growing, as well as the relative humidity. If you'll be growing in an area that has a lot of rain or moisture, your plants won't need as much water as if they were growing in a drier location. Either way, you'll need a readily available water supply. In addition, having some water barrels on standby would be very wise, provided you have that option. Monitor the weather forecast religiously and plan ahead so your crop has the best chance for success.

WHAT IS A DRIPLINE?

Once you've collected your water, one way to deliver the water to your plants is via a dripline. This low-pressure, low-volume system delivers water in a drip, spray, or stream in order to keep the roots of your plants moist.

You save water by using a dripline rather than other watering techniques because you're simply keeping the roots moist rather than soaking them.

The nice thing about driplines is you can conceal them under a layer of soil or mulch, provided the part that delivers the water is open and not clogged. Plants can also provide concealment to the dripline when they start to grow. This eliminates the special trenches required by other watering systems.

How Much Water?

t's nearly impossible to establish a set schedule without the potential worry of over- or underwatering your crop. Figuring out how much water your plants will need and how often to water them will vary greatly depending on the soil you're growing in, the local weather, and the size of your plants.

Reading your plants' leaves can indicate what's necessary. If the leaves are all drooping or drying out, your plant probably needs water. If the leaves begin to curl downward on the outside edges of the leaf, it's generally a sign of overwatering.

Overwatered

Underwatered

Collecting from Natural Water Sources

If you plan on pulling water from a natural source, such as a river or stream, first make sure you're not trespassing or breaking any laws. Once that has been established, ensure you're not drawing water from a stream that has flooded or gone dry during certain seasons. You can usually figure this out by looking for the waterline from previous years. After you collect the water, test it for contaminants and then regularly test the pH, which can fluctuate.

Using Well Water

Depending on the size of your grow, having a well installed could prove to be a wise move; however, this is pretty costly to do. If a well already exists, you're in luck. Still, you must have the water tested to ensure it's a good source for your plants, as pesticides, gases, and even heavy metals can leach into the water.

Capturing Rainwater

If possible—and after ensuring it's legal in the area you plan to grow—you can capture rainwater in barrels as a great supplemental supply for your crop. While this water should be suitable for your crops, having it tested wouldn't hurt. And as with your other potential water sources, test the pH regularly and adjust as necessary.

Watering Young Plants

When your plants are young, keep the soil moist but allow their root zone to dry out a little. This will lead to your plants developing stronger roots more quickly.

Fertilizing

In addition to sunlight, soil, and water, plants also need nutrients to grow well. While some soil mediums provide more nutrients than others, fertilizers and nutrients can help enrich your soil. While adding fertilizers and nutrients to every watering cycle isn't recommended, it's safe to do every other cycle.

Chemical vs. Natural Fertilizers

Marijuana grown outdoors with all-natural fertilizers will provide the best-tasting, smoothest-smoking buds possible. Natural fertilizers are very gentle on plants because they break down naturally in the soil, leaving no buildup or heavy metals. While it may be difficult to know the exact amount of the natural fertilizers to use, it's also harder to give a plant too many natural nutrients.

Chemical fertilizers allow you to specifically target the nutrient requirements of a plant with exact dosages. However, they can alter the taste of the buds because of salts or heavy metals that don't break down in soil. And when growing outdoors, the runoff of heavy metals can contaminate ground water in high-enough concentrations, which can build up over time. If you're using chemical fertilizers, root systems should be flushed regularly to prevent buildup of residual salts and metals on the roots.

Natural Nutrients and Additives

Natural nutrients and additives for growing marijuana are affordable, nontoxic, and sustainable. You can choose from alfalfa meal, fish emulsion and meal, and compost.

Alfalfa meal. When added to the soil, alfalfa meal corrects nitrogen deficiency by providing a natural protein. Alfalfa hay can be pressed for oil, which is then added to the soil as a liquid fertilizer.

Fish emulsion and fish meal. These are great for correcting a nitrogen deficiency efficiently. As a liquid, fish emulsion releases nitrogen quickly, while fish meal is a powder and provides a longer-term release over time with watering.

Compost. Available at any garden center or homemade, compost is a prudent additive to the soil of an outdoor grow. It provides beneficial microbes and micronutrients and acts as a buffer against disease by encouraging strong growth.

ALFALFA MEAL
The oil of alfalfa meal can be mixed with water and added to the soil as fertilizer.

Fertilizer **Application**

Fertilizers can be applied to an outdoor garden in two ways, either as a dry soil amendment (which provides nutrients over time with every watering) or as a liquid (which provides nutrients immediately).

Dry Fertilizers

Dry fertilizers can be added to the soil medium during any stage of plant growth. They're either mixed into the soil prior to planting or added to the soil around the base of an established plant. If your soil is lacking nutrients, dry fertilizers are essential to providing your roots a healthy soil medium.

Liquid Fertilizers

Like dry fertilizers, liquid fertilizers can be applied at any stage of plant growth. One advantage to using a liquid fertilizer is it can be applied topically to the leaves of your plants, providing both hydration and nutrition to the foliage (known as *foliar feeding*). Liquid fertilizers can also be applied to the soil and roots during watering.

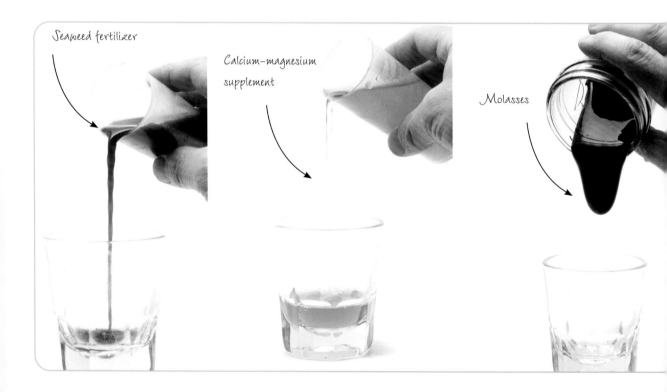

Seaweed fertilizer

Calcium-magnesium supplement

Molasses

Foliar Feeding

Foliar feeding is a supplemental and beneficial form of fertilization that should only be done during the vegetative stage. It's the fastest way to supply micronutrients to your plants and help them combat stress.

1 Add your foliar feed to water so it resembles a light tea. Follow the manufacturer's instructions to ensure you don't make the solution too strong, as even all-natural solutions can burn the leaves and stems of a plant.

2 Spray the entire plant with a fine mist, coating all the leaves and stems, top and bottom. Allow the plant to dry before watering the roots.

Dealing with the Elements

You should watch out for certain weather conditions when foliar feeding your plants. Avoid foliar feeding in the rain, as the solution will wash off. Another consideration is time of day. Don't foliar feed in direct sunlight or during the hottest part of the day, as water drops can magnify the light and cause burns. Early mornings or evenings are the best time for foliar feeding.

Foliar feeding involves spraying the leaves of your plants to supply nutrients.

The **Vegetative Stage** Outdoors

Planting outdoors allows you to use the spring and summer seasons to grow your plants to maturity. To begin in the spring, you can either start with seedlings or bring large plants outside.

How Long Is It?

Plants can begin to grow outside as early as weather allows. A plant will typically remain in the vegetative stage from early spring until late summer, when it gets closer to 12 hours of sun and 12 hours of dark (early to late August in the Northern Hemisphere). The longer your plants remain in the vegetative stage, the larger they'll get. So starting a plant indoors to get a jump-start on the size of the plant at the beginning of the season before moving them outdoors will ensure larger plants at maturity.

Protecting Your Plants Outdoors

During the vegetative stage, small plants are especially vulnerable to both weather and animals. Freezing temperatures at night or heavy hail can kill small plants in the early stage of vegetative growth. To counteract this, use small tarps or light blankets to protect small plants from these elements. In terms of predators, seedlings can easily be chewed through or trampled by any number of animals, from household pets to wild animals. To help prevent animals from interfering with outdoor plant growth, place a small fence around your plants that's anywhere from 2 to 6 feet (.5 to 1.75m) tall, depending on the predator and your plants' height.

Managing Vegetative Growth Outdoors

Healthy plants can grow several inches (cm) a day. Managing that growth allows you to influence the shape and size of the plant using pruning techniques. This is particularly important if you have limited space or concerns about public view.

Pruning can be done every few weeks, or only a few times during the vegetative stage.

1 Prune from the main stem to the tips of the branches. (Keep the fan leaves intact.) Remove any offshoots from the branches, leaving a cluster of 5 or 6 offshoots at the tip of each branch.

2 Only if you need to reduce the height or diameter, prune the tips and tops of your plants.

Keeping Your Plants **Healthy Outdoors**

Keeping your plants healthy outdoors is much more difficult than doing this indoors. Beyond the elements, you're up against wildlife, insects, and much more. Therefore, you should take certain measures to ensure your plants have the nutrition and protection they need.

Taking Care of the Soil

The soil for healthy plants should compact when squeezed but break apart easily with only a small amount of pressure. If your soil is too dense, add some perlite or clay pellets to help aerate it. This will allow the roots to grow stronger, leading to a healthier plant. Once you know your soil is healthy, you want to ensure you add the right amount of water. Too much water will cause your plants' leaves to curl inward, while too little will cause them to droop and dry out.

Providing Space to Grow

Just as fish will grow in correlation to the size of their tank, larger holes or pots equate to larger plants. To help your plants thrive, implement grow techniques (such as LST) or add a trellis. Such measures allow light and air to penetrate throughout the plants, which leads to increased yields.

Protecting Your Plants from Predators

Cannabis has many natural predators, such as ladybugs, deer, rabbits, and grasshoppers, so it's critical to put protections in place on and around your plants. Erecting cages or mesh fences around your plants helps keep out larger predators. For smaller predators, you can spray insecticide.

Ladybugs can actually protect your plants by feasting on other insects.

Safely Removing Fungus, Molds, and Bacteria

Animals aren't the only natural danger when it comes to your plants. Fungus, bacteria, and molds such as bud rot and powdery mildew can destroy your crop quickly. You can protect your plants from these issues with three particular products.

COPPER

Copper sulfate works great as an antifungal. However, it's too potent on its own. Mix it with lime so it won't harm your plants. A premixed copper sulfate and lime solution you can purchase is Bordeaux Mix. If you're applying it to young plants, dilute it heavily, and don't use it at all when your plants are in the flowering stage.

HYDROGEN PEROXIDE

Hydrogen peroxide (H_2O_2) is simply water with an additional oxygen molecule. H_2O_2 will kill most bacteria; however, that means it can kill the beneficial bacteria as well, so use it sparingly. Spraying it on infested areas is best when diluted to an H_2O_2-to-water ratio of 3:100. This will help eliminate outbreaks of algae, mold, mildew, fungi, and most bacteria.

Copper, applied during the flowering stage, protects your plants from powdery mildew.

COLLOIDAL SILVER

This product can be purchased at most garden supply stores and should be sprayed directly onto areas infected with algae, fungi, and unwanted bacteria. It's safe for humans and wildlife and won't harm your plants should you accidentally use more than the recommended amount.

Hydrogen peroxide in a diluted form protects against mold.

If your plants show signs of bud rot, remove the affected areas and scrub the rest with hydrogen peroxide.

The **Flowering Stage** Outdoors

Marijuana flowering occurs naturally outdoors because of decreasing hours of sunlight as the seasons transition from summer to autumn. Flowering generally begins in the month of August for most of North America.

Natural Light Cycles

After thriving in the natural sunlight during the spring and summer seasons, indica and sativa cannabis strains and hybrids will begin to flower when exposed to at least 12 hours of darkness. Ruderalis strains will auto-flower in summer without a change in light cycles because of the origin of their genetics in the northernmost regions of the world.

Marijuana plants shaded from early dawn or evening sun will begin to flower a few weeks sooner than plants exposed to a full day of sunlight.

Potential Diseases During the Flowering Stage

As marijuana flowers or buds mature, they become thick and dense. Increased moisture or humidity can cause bud rot, or botrytis. Cool temperatures at night along with decreased airflow through the plants can lead to powdery mildew. If buds become wet, it may be necessary to shake excess water off of them. In some cases, setting up a fan outdoors will help keep plants dry and facilitate airflow.

Removing Fan Leaves

Although it may or may not be necessary, fan leaf removal during the last week of flowering can increase airflow and light penetration for the plant.

Dead and dry leaves should always be removed from the plant and kept away from the surrounding soil. Dead leaves provide no benefit to the plant and can lead to disease or fungus and attract pests as they decay.

Healthy fan leaves—green or yellowing—can be removed a few days or weeks prior to harvest. This can facilitate airflow through the plant and help to prevent disease.

To determine when and whether or not to remove fan leaves prior to harvest, keep in mind the goal of protecting your buds from moisture and disease by providing plenty of fresh airflow. Some plants grow thicker and denser than others, and some climates facilitate molds more so than others. So judge whether it's best to do this based on the structure of your plants as well as the climate.

Light Deprivation Method

Reducing light exposure to 12 hours or less will cause marijuana plants to flower. Light deprivation techniques are employed outdoors to harvest early and in some cases allow for more than one harvest per season.

Techniques

Depending on the size of your plants, several techniques can be employed to deprive your plants of light, from small boxes or garbage bags that block out the sun; to tarps, tents, and hoop houses; to large grow structures with blackout shades.

Timing

A savvy outdoor grower can harvest twice a year, especially in regions where the seasons are conducive to an early spring and late fall. For instance, planting in April allows for a harvest in July (if using light deprivation techniques), while a second planting left to mature all summer (or planted in July) can be harvested in October using natural light.

Scheduling

Once you've decided on a light deprivation technique, it's important to schedule the time your plants will receive light—for example, 7 A.M. to 7 P.M. With a lighting schedule in place, you must follow it religiously. A dramatic change in the light deprivation schedule will stress a plant, potentially stunting its growth or causing it to turn into a hermaphrodite and produce seeds.

Building a Hoop House

One technique for depriving your plants of light is a hoop house. Constructing a hoop house is simple and inexpensive.

WHAT YOU NEED

6 to 8 pieces 2- to 2½-foot-long (.5 to .75m) rebar or similar staking material
Clamps and bolts
6 to 8 lengths ½- to 2-inch (1.25 to 5cm) PVC pipe
Plastic sheet, enough to cover the tops, sides, and ends of the hoop house
Light-blocking material, such as a tarp, to cover the plastic sheet

1 Face two rows of stakes directly across from one another, approximately 6 feet (1.75m) apart per row. Drive stakes into the ground, leaving them about 1 foot (30.5cm) or so above the ground. Bolt clamps to the wood where you'll place the pipes.

2 Use PVC pipe to form an arch between the two rows, sliding the end of the pipe over the top of each of the stakes positioned across from one another and securing them in the clamps. (The length of the PVC pipe and the distance between the two rows determine the height and arc of the structure.)

3 Place the plastic sheet followed by the light-blocking material over the top and both ends of the PVC hoops. This will completely block out the sunlight from your plants during the scheduled light deprivation time. When that time is complete, remove the tarp for 12 hours of sunlight during the day.

Harvesting Outdoors

Once you've made it through the outdoor growing season, you can harvest your marijuana plants. So what do you need to do? How should you prepare? Knowing when and how to harvest is key, as everything is riding on this process.

Determining When to Harvest

Just as with harvesting indoors, your plants will show telltale signs of being ready to harvest. When ready, your plants' leaves will begin to change color, while the trichomes will turn from an opaque to amber color. Because not all of your plants will be ready at the same time, weather permitting, you can harvest the tops and leave the bottoms for a few more days. This will allow the other buds to utilize resources they were previously denied until those are ready for harvest, too.

TRICHOMES
These can indicate when to harvest your plants. The trichomes will appear opaque to amber at maturity.

Preparing for Harvest Time

You should make certain preparations before you harvest to ensure a healthy and successful crop. It is imperative that you dry your plants indoors after harvest, as the elements, pests, and wildlife will most definitely take their toll on your finished crop. Therefore, you must decide where you're going to dry and trim your plants and have it prepared prior to taking the plants down. Have a place to hang your plants that will allow them to dry in a room with conditions conducive to this step. Odor control, temperature, air movement, and humidity control will need to be addressed in your drying room prior to starting the drying process.

When it comes to the plants themselves, flush your plants 1 or 2 weeks before your planned harvest in order to remove any toxins from the fertilizers. A few days before harvesting the buds, you can allow your plants to droop a little so they're less moist when it's time to dry them. While you're managing these steps, keep an eye on the weather conditions for the time of your anticipated harvest so you're not planning to harvest on a rainy day.

Documenting Your Harvest

A good grower always takes notes on the entire grow cycle of each crop. When you're taking down your crop in sections, document and track your movements. This also allows you to better prepare for the next harvest. For instance, buds will lose 70 to 75 percent of their weight from harvest to cure—that's a huge difference. Documenting these types of changes in your current harvest helps you know what to expect with your next one, making your process more successful. The more information you have, the better you can assess your skills and growing methods.

How to Harvest Outdoors

Once you've decided when you're going to harvest and have made the necessary preparations, you can begin harvesting. Don't take down more than you can handle at one time (for instance, if you plan to trim first, cut down more than you can trim before the plants start drying), and start early in the day. Using pruning scissors or shears, cut off large clusters of buds located on the tips of the plant branches. Smaller buds that remain further down along the length of the branches can be trimmed off, or the branches can be removed entirely prior to drying.

Remove the fan leaves, if time permits, and hang your plants upside down in your drying room. Fan leaf removal is much easier during harvesting, as drying makes it more difficult to completely remove them. Because fan leaves left on a bud contribute to a lesser-quality aroma, taking care of it now is better in the long run.

Part 5
Extending Growing Seasons with Structures

You can construct several types of structures for your personal-use cannabis garden to extend your growing seasons. These can range from inexpensive temporary structures (such as simple hoop houses) to permanent and more-expensive structures (such as a greenhouse). In this part, I discuss the benefits of structures, how to build them, and ways to make the conditions in them ideal for your plants.

Why Grow in Structures?

Growing in structures allows you to experience the best of both indoor and outdoor growing. Outdoor grow structures provide your plants with natural sunlight and airflow, while also offering climate controls and protection from extreme environmental conditions.

Extending Your Growing Season with Structures

The primary purpose of growing in structures may be to protect plants from extreme weather, but depending on the type and quality of your structure, you may be able to grow your personal-use cannabis plants all year long, regardless of the season.

Creating a Better Environment for Growing

Grow structures work by allowing natural sunlight to penetrate the outer covering of the structure. In addition to providing protection for your plants from the elements, they retain heat and humidity. These hot and humid conditions require growers to provide temperature control and ventilation for the structures throughout the year to compensate. (These demands vary based on your climate and season.)

To help with temperature and humidity, allow air to flow through the grow structure when weather allows. You can also use fans, heaters, or coolers (swamp coolers or air conditioning), if necessary.

A Structure for Every Type of Grower

Cannabis plant growing structures vary significantly. From simple and inexpensive structures to technical and costly ones, you'll find growing structures that allow you to dig your plants into the ground or install a state-of-the-art hydroponic growing system.

On the cheaper side, some provide little more than basic shelter from rain, have dirt floors, and have coverings that can easily be removed to allow an open-air experience for your plants. Other, more expensive and solid structures have concrete floors, foundations, glass walls, a convertible roof, climate controls, electricity, and even plumbing. So whatever your needs and climate, you can find or create a grow structure that suits them.

GREENHOUSE THERMOMETER
With a greenhouse thermometer, you can monitor the temperature in your grow structure to ensure optimal growing conditions for your plants.

Types of **Grow Structures**

The type of grow structure you decide on will depend on your budget, your climate, and the goals you wish to accomplish. Temporary structures are the most cost effective and will extend your growing season slightly, offering protection from the elements throughout the growing season. Permanent structures are more expensive but will extend your growing season (if properly heated) and protect your plants year-round.

Cold Frames

Cold frame grow structures are generally temporary and inexpensive to build. Single walled in either plastic or glass, they come in all shapes and sizes and have no temperature controls. Cold frames are effective in extending growing seasons for a few weeks in the spring and fall by preventing frost from forming on your plants while keeping the ground a few degrees warmer. Simple cold frames can cost less than $100 to $200 for materials.

Hoop Houses

Hoop houses are one type of cold frame structure. Inexpensive and simple to build, they typically consist of PVC pipes that are secured at each end into the ground by rebar or a wood frame to form a "hoop." Clear or opaque plastic is then stretched over the hoops for a covering. The ends can either be left open or wrapped in a plastic covering as well. You can use a fan in either end of hoop houses to facilitate ventilation and to help prevent humidity. You can also employ a light deprivation technique with hoop houses by using a tarp or covering that blocks light.

Hot Frames (a.k.a. Greenhouses)

Hot frames or greenhouses provide temperature and climate controls that allow you to maintain an ideal temperature above 55°F (13°C) and below 90°F (32°C). They have adequate ventilation and are capable of being open to the air or completely sealed from the elements. Hot frames can be walled in with plastic or glass and equipped with either manual or automatic light deprivation equipment. They're often double walled for colder climates, especially when used for year-round growing. Hot frames can cost from around $1,000 to several thousand dollars, depending on the quality and options you decide on.

Controlling Climate and Environment

Controlling climate is essential to the healthy growing of your personal-use marijuana plants. The extent to which you can control the environment in either a cold frame or hot frame grow structure depends on several factors, including the quality of construction and materials.

Active Climate Control

Active climate control systems can be built into hot frame grow structures. Active climate control equipment includes swamp coolers, ventilation fans, air conditioning units, and heaters. Hot frames must be built properly to ensure both efficient climate control and safety in order to facilitate the incorporation of mechanical equipment.

Passive Climate Control

Passive climate control is basically the classic combination of sun and fresh air. The sun warms the air and ground inside cold frame or hot frame grow structures, while fresh air circulates naturally throughout them. In cold frame grow structures, air can passively circulate by uncovering openings in the sides, ends, or top of the structure. In hot frame structures, doors, walls, skylights, and sometimes entire roofs can be open or lifted to allow the fresh air in.

Supplemental Lighting

In addition to active climate control equipment, hot frames can also be equipped with supplemental lighting (with HID bulbs) that allows you to provide light to the environment during periods of darkness in the early morning or evening hours or during overcast days. (Supplemental lighting should only be used in hot frames with properly installed wiring.)

Light Deprivation

You can employ light deprivation manually or automatically. Manual deprivation techniques include inexpensive light-blocking materials, such as colored plastic or tarps. These can be placed and removed by hand when required—either draped over cold frames or pulled like curtains across the sides and ceiling inside hot frames.

Automatic light deprivation equipment is a more expensive, albeit efficient, option that allows for total light blockage at the touch of a button as needed. This equipment, controlled by electric motors and electronic timers, opens and closes the blackout materials for the ceiling and walls, allowing for total darkness in bright sunlight.

Choosing the **Right Grow Structure**

To choose the right grow structure for your application, you should first know what your goals are (such as protection from the elements or year-round growing), in addition to the climate conditions you'll be growing in. Grow structure options range from very basic to cold frame structures to state-of-the-art hot frame structures.

Simple and Inexpensive

The simplest grow structures are generally cold frame and can be constructed easily as a home do-it-yourself project with inexpensive building materials, such as wood or plastic. However, simple hot frame grow structures can also be constructed relatively inexpensively. This construction may include dirt floors and very few or no pieces of climate control equipment. Simple cold frame structures can cost as little as a few hundred dollars, while simple hot frame structures will cost around $1,000.

Determining the Size of Your Structure

The size of your grow structure should be large enough for the number and size of marijuana plants for your personal use. Sizes can range from less than 100 square feet (30.5m) or less to several hundred square feet (m) or larger. Additionally, your personal grow space may be incorporated into a larger growing structure. Just remember to check your local laws to ensure the size of your structure is legal.

More Complex and Expensive

You can design state-of-the-art grow structures that are capable of environmental control in the coldest or warmest of extreme climates. Hot frames are the most complex structures and can include the full automation of climate controls for temperature (to include heated floors) and humidity, as well as automated timers to operate supplemental lighting, light deprivation, airflow, and watering systems. The most complex structures are constructed from glass, concrete, and steel, and are completely wired for electricity and plumbed for water use and drainage.

A small, fully equipped hot frame capable of year-round growing in extreme climates can cost at least $10,000, while large commercial units can cost up to $100,000.

Building a **Grow Structure**

Construction of your grow structure can range from an easy, afternoon do-it-yourself project for a small, simple cold frame to hiring contractors for the construction of a permanent, year-round hot frame.

Zoning Requirements

Whichever grow structure you decide to construct, you should first check your local city or county zoning requirements. Depending on your municipality, zoning may or may not allow the construction of either cold frame or hot frame structures. In addition to zoning requirements, your neighborhood may have covenants, conditions, and restrictions on what you're allowed to construct on your property.

Building Permits

Building permits generally aren't required for cold frame grow structures in most municipalities where they're allowed, due to their temporary nature. Hot frame structures generally require a permit because of the permanent nature of building on a foundation and the technical aspects of wiring and/or plumbing. However, some small greenhouses come as ready-to-assemble kits and may not require a foundation or a permit.

Always check with your local building officials before beginning construction on any type of grow structure—either cold frame or hot frame—in order to avoid costly fines or the possible removal of your structure.

Choosing a Location

Choose a location that's logistically correct in terms of convenience and efficiency of land use, as well as a location that's aesthetically pleasing to your landscape and your neighbors. Whenever possible, locate your grow structure in the same place you would place your outdoor garden. Ideally, your structure would be on a level surface in an open area away from shade to obtain as much sunlight as possible.

After you've determined where you would like your grow structure to be located, be sure it's allowed by your local neighborhood CC&Rs (covenants, conditions, and restrictions) and municipal zoning and obtain any of the required permits as necessary.

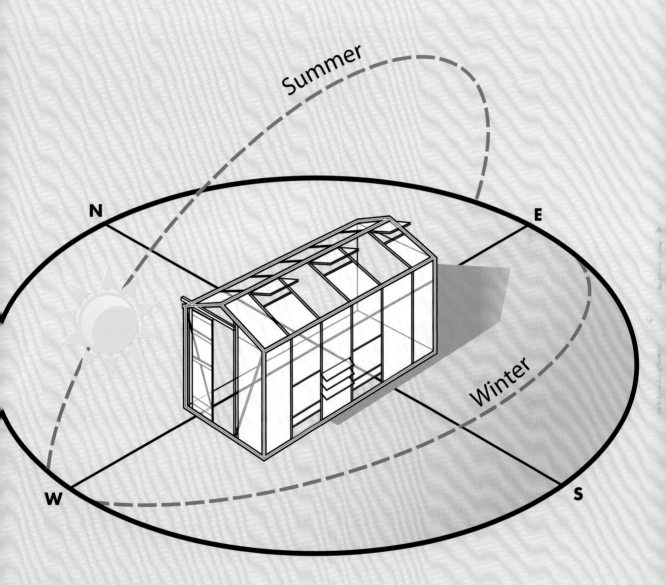

POSITIONING YOUR GROW STRUCTURE

In this image, the grow structure is positioned ideally in regard to the arc of the sun, allowing full, natural sunlight from dawn to dusk in both the summer and winter seasons.

Part 6
Breeding

Breeding is a technique that shouldn't be tried until you achieve expert grower status. It combines the genetics of one plant with the genetics of another to create a new plant with its own unique genetic makeup. This truly rewarding hobby allows you to intentionally breed plants to add or remove certain genetic qualities. In this final part, I tell you all you need to know about breeding cannabis so you can cultivate a crop to your preferred tastes.

The Goal of **Breeding**

Breeding allows you to customize your strain of marijuana. The goal of breeding is to create a genetically stable and superior plant with certain desired characteristics, or traits, from the combination of one plant with another.

Crossing

Crossing—also known as *cross-pollination* and *cross-fertilization*—is the process by which two strains are bred to each other by adding the pollen of a male flower to the surface of a female flower. When done correctly, you should be able to produce seeds that grow into plants that express the desired characteristics of each parent. The unique expression of the traits given to a particular strain of cannabis is known as the plant *phenotype*.

Before you begin crossing plants, you have to determine what traits you want to be present in your future personal-use cannabis plants. Do you wish to breed a hybrid of two genetically different marijuana plants? If so, which traits from which parents do you want to keep for future generations? Intentionally selecting the genetic traits you want allows you to avoid the random hybridization found in nature, thereby giving you control over the final outcomes.

Selective Genetics

Breeding and selecting only the best plants for your desired goal over generations and generations will ultimately result in your ideal plant. So how do you do this? When selecting your plants for breeding, look for the hereditary expressions of the genetic traits you desire in both the male and female parents.

Some of the desirable traits that you may choose to breed into your plants may include, but are not limited to, the following:

Hardiness. This refers to resistance to environmental stressors like heat, cold, drought, wind, and so on.

Disease resistance. Certain plants are less susceptible to molds and fungus than others.

Pest resistance. Look for plants that have evolved a natural tendency to resist pests.

Plant size and yield. Height, width, growth pattern, and overall yield by weight vary among strains.

Adaptability. Cannabis plant genetics have derived from plants that have originally adapted to their particular climates in various parts of the world.

Flowering times. The genetics of a plant dictate how long it takes for the plant to mature and flower.

Breeding Based on Environment

While environmental conditions can enhance certain traits or the performance of marijuana plants, these factors aren't necessarily expressed through heredity.

Selecting **Parents**

Selecting your initial cannabis parent plants for breeding determines what traits or ranges of traits your offspring can possibly have. For optimum results, select two varieties to cross together that are dissimilar, such as an indica male and a sativa female. This will usually lead to more novel trait combinations in the resulting offspring.

Selecting Female Parents

Choosing female parents is a relatively easy process. You simply look for a female plant whose traits you desire in your future cannabis strain. These traits might include growth pattern, leaf shape, plant size, pest and disease resistance, and a variety of other characteristics.

Once you decide what you want, you need to know the quality of the buds from your desired female parent; after all, you want offspring that have both your desired traits and quality buds. So unless you already have an idea of what the bud from a variety is like, take time to flower out a clone and test the buds before using a particular variety as a female parent.

FEMALE
CANNABIS PLANT

Flowers

Breeding Warning

As with other parts of the process, having healthy starting plants is important. If you select weak parents that have been improperly stressed before crossing, their offspring might suffer from degraded genetics, resulting in decreased yield, vigor, and potency.

Selecting Male Parents

Choosing males is one of the more difficult tasks in breeding cannabis plants, as you can't simply flower a clone and test the bud characteristics. However, there are certain things you can do to find out what traits a male plant might have, as well as its potency.

WAYS TO CHOOSE MALE PLANTS

Often, the primary method of choosing a male parent involves simply observing the overall growth and vigor of it. To know the bud characteristics a particular male might pass on, you can flower and test the buds of a female plant of the same strain.

TESTING FOR POTENCY

Testing laboratories can run inexpensive tests on male plant material to determine if a particular male might produce offspring with high THC or CBD. However, one old-school method for testing male potency is to have someone with a very low tolerance for the effects of cannabis smoke the pollen from a mature male flower. Because more potent males will produce trace amounts of THC with their pollen, a tester with low tolerance should be able to feel the effects.

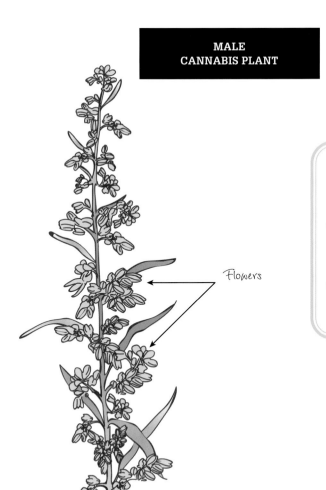

**MALE
CANNABIS PLANT**

Flowers

Early Flowering Males

Many breeders believe that by selecting early flowering males, you're selecting for other hemplike characteristics. This is a bad thing, and breeders will discard their quickest-flowering males as potential male parent plants.

Collecting Pollen from Male Plants

After you've chosen your parent plants, you'll collect pollen from the mature male plant and pollinate the mature female plant. The following walks you through how this process works.

Before Collection

Flowering males should always be grown away from flowering female plants you're intending to harvest for bud. Otherwise, pollen will be dispersed by a gentle breeze or oscillating fan, causing your females to set seed and ruining your crop of buds.

Neighboring Crops

If you have any neighbors growing marijuana, care should be taken to quarantine your males from their females as well, so you don't accidently ruin their crop via pollination.

When to Collect Pollen

Correctly selecting the right time to collect your pollen will allow you to obtain many viable, healthy seeds. Cannabis pollen should be collected from healthy males during the optimum moment of pollen sac maturity. Male flowers show they have mature pollen as the first pollen sacs split open, revealing three hanging anthers (which resemble little bananas). Inside these anthers is the pollen, which should be a very faint yellow color, and resemble baking flour in consistency.

If the pollen sacs are milky and white when broken open, the male flowers are not yet fully mature. If pollen collection is delayed too long, the anthers will burst open and drop their pollen on the ground. Therefore, it's best to check your male plants for maturity each day. Because not all flowers on a plant will mature at the same time, pollen should be harvested several times from the different flowers as they mature on each male plant. And while you might not collect much pollen from a male plant, a little will go a long way in pollinating female plants.

Cautionary Notes About Pollen Collection

When you collect pollen, which at times can be impossible to see, you'll find it has a tendency to fly everywhere. It can get on your hands, as well as stuck on your hair, clothing, and even arm hair! If it has stuck to you and you then go work around females in another room, you could inadvertently pollinate plants you don't want to. To keep the pollen as under control as possible, avoid any kind of breeze in the room. You should also make sure to bathe and change your clothes (and wash them) immediately after collecting pollen.

Plus, bear in mind that pollen can get into lighting fixtures, the minute fissures in tools and equipment, and the fibers in carpet, among other things. So before you expose female plants to anything that's been in contact with pollen, clean it thoroughly.

How to Collect Pollen

You can experiment with methods to collect mature pollen. The following is a simple method:

1 Place a hard surface (such as a mirror) around the base of the mature male flower clusters.

2 Lightly shake or tap the male plant so the pollen goes onto the hard surface. (You can also use a small paintbrush to lightly dust the pollen out of the pollen sacs and onto the surface.) Because the sacs will keep spilling pollen, you can keep collecting for up to 3 weeks.

3 Gently scrape the pollen into a vacuum bag or jar, and store it in the freezer. Correctly preserved mature pollen can be stored for up to 6 months.

How to **Pollinate**

Pollinating should typically occur about 3 or 4 weeks after the female plants are flowered. Before you begin this process, make sure your female plants are at the correct maturity to receive pollen. If you pollinate your females too early or too late, the plant won't produce as many healthy and viable seeds. Maturity can best be judged by observing the female flowers. The flowers should be fully formed, and all the pistils should be white.

Active Pollination

To actively pollinate female plants, use a soft-bristled paintbrush to apply the pollen that was properly collected and stored at an earlier time. This is accomplished by dipping the paintbrush into the pollen, and then lightly brushing the paintbrush over the calyx and pistils of the female flower. After pollinating the flowers of the receptive female, allow the female plant to mature to the point that seeds form.

Avoiding Contamination

To avoid contaminating other flowering plants, you should perform the pollination process in a separate area. Also, be sure you clean your clothes and shower after leaving a pollination room and before moving to any other flowering-crop environment.

Passive Pollination

Often, pollinating female cannabis plants can be accomplished without collecting pollen at all. Instead, you can put females into flower 2 or 3 weeks sooner than males and allow the males and females to flower in the same room. Airflow from the fans of an indoor grow room will blow the pollen from the flowering males onto the receptive female flowers.

You can also shake a pollinating male cannabis plant above the female plant to passively pollinate. In this method, the male pollen sacs—which should be ready to rupture—spill their pollen freely on the receptive female plants when you shake them. When doing this, take care to pollinate only those females you have chosen for breeding, as the smallest amount of pollen from any male can cause any receptive female to germinate seeds.

COLLECTING POLLEN
After you apply pollen with a paintbrush, you can collect excess pollen in a tube.

Collecting **Fertile Seeds**

Once you've pollinated your female plants and allowed them to mature, you can collect their seeds. Keeping your plants healthy as seeds develop will help pollinated females produce many viable seeds. Be sure to properly care for your pollinated females by giving them adequate heat, light, water, and anything else you might give flowering plants as they mature.

Ensuring Seeds Are Mature and Healthy

Before you begin collecting seeds, remove and test a few seeds from the bottom flower clusters of a plant to see whether they're mature (as a plant will fully mature starting from the top of the plant).

The following guidelines can help you judge your seeds' maturity:

Mature and viable. Seeds that are mature and viable are full and dense, with a hard outer brown layer. Often, this brown outer layer is mottled or spotted.

Not mature. If seeds aren't fully mature, they'll be soft and milky.

Unhealthy and unviable. Seeds that aren't healthy and viable will look smaller and deflated compared to mature seeds, with an outer layer that's bleached white.

Collecting Seeds

Here's how to harvest and collect seeds:

1 Cut the mature plant whole at the base of the stem and hang it upside-down to dry.

2 When the plant is dry, take each dry, seedy bud and rub it back and forth between your hands over a work surface to separate the seeds from the plant material.

3 Gather the seeds by screening out the plant material, hand picking out the seeds, or a combination of methods.

Storing Your Seeds

Store viable seeds in an airtight container and keep those containers in a cool, dry place. If properly stored, your cannabis seeds can remain viable for several years. And if you've harvested and stored enough seeds from a specific cross-section of plants, you can use those as a test run for germination. Simply attempt to grow a small lot of the seeds and see how many of them germinate and grow.

Stabilizing **Desired Traits**

Stabilization refers to a decrease in your plants' genetic variance when natural selection favors an average phenotype (characteristic) and selects against extreme variations. Gathering information in order to stabilize the desired traits for your plants can happen through many methods, from the empirical and tactile to more mathematical and cerebral.

First-Generation Offspring: The Crosses

The plants grown from the seeds gathered from the first cross between your chosen parents will be genetically similar, resulting in a small diversity of plants to choose from. Diversity can be observed in how a plant grows; what its flowers are like; how it smells, tastes, and feels; how resistant it is to problems; and its water and nutrient requirements. However, selection should not be performed during the first generation of offspring, as there isn't enough diversity to reliably select from. Instead, healthy males and females from this first generation should be crossed.

Second- and Later-Generation Offspring: Stabilization

In the second generation of offspring, male and female plants will express the greatest variety of observable traits from two original parent plants. This is the generation where selection for traits will be the most intense. Some breeding techniques might select very intensely in this generation (by only choosing very few parent plants to breed), and some breeding techniques might select less intensely (by allowing most of these parent plants to breed). Whatever the case, in this and subsequent generations, you can choose male plants to collect pollen from and female plants to pollinate that have the traits you desire in your stable strain.

Selecting Carefully

Think through what you are selecting for carefully. Because traits can be linked together, intense selection of a desired trait at the wrong time might negatively influence another desired trait. For instance, something like height or leaf size might be related to the cannabinoid production in each offspring plant, so keeping all the tall plants without testing potency first might produce a less potent strain once your stabilization is finished.

You should collect enough second-generation seeds (and seeds of each following generation) that you have a good deal of diversity to select from. While it's wise to germinate at least 100 seeds in each generation, most genetic diversity can be observed with 30 or more seeds in each generation. Over multiple generations, where the selected male and female offspring are crossed together, the traits of the offspring will more closely resemble each other. This is the essence of stabilizing a strain. It will take between 7 and 10 generations to stabilize a strain.

Feminization

By using feminization (or by simply releasing clones from the first generation of new crosses), you can ignore strain stabilization altogether. Feminization is an advanced technique used by breeders when they've found a truly exceptional female plant before the strain has been stabilized. Often, feminization is useful because an unstable strain can have valuable traits (such as yield and potency) that outperform both original parents and also outperform any stable version of the same strain. Therefore, it may sometimes be useful to immediately stabilize a rare combination of traits in seed form. However, some breeders avoid feminization altogether because they believe it leads to a less potent and less flavorful final stabilized strain.

Backcrossing

Backcrossing is another advanced technique used by breeders in which they have an original strain they value but would like to introduce a trait or set of traits found in a different strain. Using this technique, you would select clones from a female and pollinate them with a desired male without killing off the original mother (or ensuring that a nonflowering clone of the mother is always preserved). The seed from this first cross would be germinated, the original female mother would be cloned, and the male seedlings and female clones would be flowered. Only the seeds from the desired males should supply the pollen for each new round of clones from the original female mother. As each new generation is crossed, the strain will stabilize back to the original mother's characteristics while bringing along the desired characteristics from the male parent strain.

Stabilizing Cannabinoid and Terpene Profiles

If you're looking to select plants for stabilization whose clones have the most desired cannabinoid and terpene profiles, you might choose to take clones from the females in each generation and flower these clones before flowering the potential breeding females. You can then either test the cannabinoids and terpenes of each clone in a lab chromatograph or simply smoke the flower from these clones and record your results.

Glossary

aeroponic   A type of hydroponic growing in which water is pumped from a reservoir and sprayed on the exposed roots in a container.

auto-flower   Sativa or indica plants that have been cross-bred with ruderalis plants to create a strain with a 70- to 90-day clone-to-harvest grow cycle.

British thermal units (BTU)   Measures the output produced by an air conditioner.

broad leaves   Large leaves that serve as photosynthetic factories for the production of sugars and other necessary growth substances. *See also* fan leaves, water leaves, and sun leaves.

calyx   A collection of sepals that forms a protective layer around a plant's bud.

closed growing environment (CGE)   A completely airtight indoor grow room designed to run without outside interference or interaction with the elements outside the room.

curing   The process of placing dried buds in airtight jars and exposing them to air for a few moments a day. Cannabis buds are properly cured when they feel slightly sticky and moist but not brittle.

deep water culture (DWC)   A type of hydroponic growing system in which water is pushed up through an air pump while the roots dangle in the water. This allows for constant feeding of your plants.

double-ended (DE) bulb   A high-power sodium bulb that's lit on both ends, allowing a much brighter light.

ebb and flow   A type of timed hydroponic growing system in which water is pumped from a reservoir into a table where plants are growing. After a set amount of time has elapsed, the pump shuts off until the next cycle.

fan leaves   Large leaves that serve as photosynthetic factories for the production of sugars and other necessary growth substances. *See also* water leaves, sun leaves, and broad leaves.

fulvic acid   A soil supplement used in conjunction with humic acid to help break down nutrients in the soil and make them more available to plants.

Fan leaves

germination   The process of growing a plant from a seed.

grow tent   A small portable structure that provides a self-contained room in which to grow indoors.

hardening   Also known as *hardening off,* the process of conditioning a plant to acclimate to harsher conditions during the transition from the vegetative stage to the flowering stage. This can be done for both indoor and outdoor growing.

hardiness map   The standard by which gardeners and growers can determine which plants are most likely to thrive at a location. The map is based on the average annual minimum winter temperature divided into 10-degree zones.

high-pressure sodium (HPS) bulb   A high-intensity discharge bulb primarily used during the flowering stage of plants.

humic acid   A soil supplement used in conjunction with fulvic acid to help break down nutrients in the soil and make them more available to plants.

humidity pack   A humidity pack maintains a specific humidity within a cannabis storage container by drawing moisture out of the air or adding moisture to the air.

hydroponic   System developed that utilizes water to grow plants. There are several different variations of hydroponic systems, including aeroponics, ebb and flow, drip systems, nutrient film technique (NFT), and wick.

hydroton   A grow medium made from clay that's used in most hydroponic applications or as a soil additive to help with aeration.

Indica nug

hygrometer   An instrument that's used to measure humidity.

indica   One of the three original species of cannabis. *See also* sativa and ruderalis.

internodal   The space in between nodes on the branch of a plant.

K   The chemical symbol for potassium, one of the three macronutrients required in fertilizer to feed cannabis plants.

Kelvin An absolute scale of temperature in which the degree intervals are equal to those of the Celsius scale and in which absolute zero is 0 Kelvin and the boiling point of water is approximately 273 Kelvin.

light-emitting diode (LED) A semiconductor diode that emits light when conducting current and is used in electronic displays, indoor and outdoor lighting, and so on.

lumens The unit of luminous flux used to measure the strength of a lightbulb.

metal halide (MH) A type of bulb used to grow plants, mostly in the vegetative stage.

node A joint or knot in a plant's stem that normally bears a leaf. This is where the plant branches out between the internodal sections.

NPK Stands for "nitrogen, phosphorus, and potassium," the important elements in a fertilizer or soil amendment. Nitrogen is responsible for strong stem and foliage growth, phosphorus aids in healthy root growth and flower and seed production, and potassium is responsible for improving overall health and disease resistance.

nug Slang for a high-quality marijuana bud.

nutes A slang term for nutrients used to feed cannabis plants.

parts per million (PPM) A value that's often seen in relation to the concentration of minerals in the water given to plants.

pistil The ovule-bearing or seed-bearing female organ of a flower, consisting of the ovary, style, and stigma.

Powdery mildew

powdery mildew   A disease characterized by the yellowing and death of foliage and a white, mealy growth of fungus on the surface of plants.

preen   The technique of removing unproductive or dying leaves in an effort to promote the overall growth of a plant.

processing   Trimming, drying, and curing usable cannabis; cooking with marijuana; or making hash for consumption.

production technique   The act of producing or growing marijuana in various stages, from seed to flower.

propagate   To multiply by any process of natural reproduction. Cloning a plant is considered propagation.

prune   To cut, clear, or get rid of superfluous or undesired twigs, branches, or roots from a plant.

relative humidity (RH)   The amount of moisture in the air measured as a percentage.

reverse osmosis   A water purification technology that uses a semipermeable membrane to remove larger particles from drinking water. Reverse osmosis filters are used in hydroponic systems to encourage health and growth in plants.

rock wool   Spun rock used as a grow medium in most hydroponic systems.

root rot   A symptom or phase of many diseases of plants, characterized by discoloration and decay of the roots. This is usually brought on by overwatering a plant.

ruderalis   One of the three main species of cannabis. *See also* indica and sativa.

Sativa nug

sativa   One of the three main species of cannabis. *See also* indica and ruderalis.

scrogging   Also known as the *screen of green* method of growing, the technique of putting a trellis net or screen over your crop with the intent of training the plants to grow up and through the covering.

sepals   The parts of the calyx of a flower, enclosing the petals. These are typically green and leaflike in appearance.

silica   A chemical compound that's the main constituent of most of Earth's rocks. Silica can improve the overall health of a plant by reinforcing and strengthening the epidermal cell walls. It's stored primarily in the spaces between the cellulose micelles.

sugar leaves   The actual leaves growing out from within the bud of a plant. Sugar leaves are coated in the resinous glands of the plant and are usually used to process hash and other concentrates.

sun leaves   Large leaves that serve as photosynthetic factories for the production of sugars and other necessary growth substances. *See also* fan leaves, water leaves, and broad leaves.

super cropping   Describes a variety of techniques used to increase the smokable yield of cannabis crops. All super cropping techniques basically involve restricting or damaging the plant in a self-repairable manner to promote accelerated growth activity.

T5   A fluorescent fixture that takes a T5 bulb. More powerful than a T8, the T5 outputs more light but also more heat.

T8   A fluorescent fixture that takes a T8 bulb. T8s allow the plant to touch the bulb when using a "daylight" bulb, about 6,500 K.

THC   Stands for *tetrahydrocannabinol,* the active chemical in cannabis responsible for a user's high.

topping   A form of super cropping in which the top stem of a plant is removed so the resources are then diverted to the next tier of stems, allowing them to flourish.

trichomes   The resin glands that form on the buds of a cannabis plant. They're what makes the bud sticky and tend to have the majority of THC.

usable marijuana   The part of the female cannabis plant, including the flowers or buds, that's processed for consumption.

vapor pressure deficit (VPD)   A way to measure the climate of a grow structure. VPD can be used to evaluate the disease threat, condensation potential, and irrigation needs of a structure's crop.

veg   A slang term referring to the vegetative stage of cultivation.

water leaves   Large leaves that serve as photosynthetic factories for the production of sugars and other necessary growth substances. *See also* fan leaves, sun leaves, and broad leaves.

Trichomes

Resources

The following websites, journals, and books are sources I used for information in this book. Feel free to explore them for more information.

Websites

These websites cover many different topics related to the marijuana-growing process.

BULBS
leafly.com/news/growing/its-time-to-upgrade-your-grow-lights-to-led

CALYX
thefreedictionary.com/calyx

CLIMATE AND MARIJUANA
en.wikipedia.org/wiki/Hemp

COLLOIDAL SILVER
naturalnews.com/038579_colloidal_silver_healing_water.html#

GROW TENTS
buyersguide.org/grow-tent

leafly.com/learn/growing/buyers-guide/indoor-grow-tents

HARDINESS MAP
planthardiness.ars.usda.gov

HARVESTING OUTDOORS
ilovegrowingmarijuana.com/growing/harvesting

HUMIC AND FULVIC ACIDS
wellnessmama.com/25300/what-is-fulvic-acid

HUMIDITY PACKS
weedmaps.com/news/2023/01/best-humidity-packs-for-weed

KEEPING OUTDOOR PLANTS HEALTHY
happysprout.com/inspiration/summer-heat-and-plants

marijuanagrowershq.com/how-to-harvest-marijuana

KELVIN
dictionary.reference.com/browse/kelvin?&o=100074&s=t

LED
dictionary.reference.com/browse/led?s=t

LIGHTING
just4growers.com/stream/grow-lights/air-cooled-reflectors%E2%80%94do-you-believe.aspx

NODE
dictionary.reference.com/browse/node?s=t

NPK
organicgardening.about.com/od/organicgardeningglossary/g/NPK.htm

PISTIL
dictionary.reference.com/browse/pistil

PM (POWDERY MILDEW)
dictionary.com/browse/powdery-mildew

PPM
rapidtables.com/math/number/PPM.htm#definition

PROPAGATE
dictionary.reference.com/browse/propagate

PRUNE
dictionary.reference.com/browse/prune?s=t

SETTING UP A HYDROPONIC SYSTEM
helpfulgardener.com/forum/viewtopic.php?t=32071

SILICA
dictionary.com/browse/silica

cannabis.info/us/abc/30007838-silica-auxins-and-cytokinins-a-trio-of-cannabis-plant-additives

SPECIES

Afghanica
theleafonline.com/c/science/2015/01/indica-sativa-ruderalis-get-wrong

Ruderalis
leafly.com/news/cannabis-101/what-is-cannabis-ruderalis

americanbar.org/groups/senior_lawyers/publications/voice_of_experience/2021/voice-of-experience--june-2021/cannabis-101-history-and-terminology

Sativa
leafly.com/news/cannabis-101/sativa-indica-and-hybrid-differences-between-cannabis-types

STRAINS

9 Pound Hammer
leafly.com/indica/9-pound-hammer

allbud.com/marijuana-strains/indica-dominant-hybrid/9-pound-hammer

Afghani
sensiseeds.com/en/cannabis-seeds/sensi-seeds/afghani-1

leafly.com/indica/afghani

Afghani OG
en.seedfinder.eu/strain-info/Afghani_OG/Apothecary_Genetics/

Berry White
allbud.com/marijuana-strains/indica/berry-white

leafly.com/strains/berry-white

Blue Hawaiian
allbud.com/marijuana-strains/hybrid/blue-hawaii

en.seedfinder.eu/strain-info/Blue_Hawaiian/Jordan_of_the_Islands

Blue Lavender
en.seedfinder.eu/strain-info/Blue_Lavender/Exotic_Genetix

Blueberry
leafly.com/indica/blueberry

allbud.com/marijuana-strains/hybrid/blueberry

Cactus
leafly.com/indica/cactus

CBD Critical Cure
leafly.com/indica/cbd-critical-cure

allbud.com/products/breeze-botanicals/flower/35484/cbd-critical-cure

Chernobyl
leafly.com/hybrid/chernobyl

allbud.com/marijuana-strains/sativa-dominant-hybrid/chernobyl

Day Tripper
allbud.com/marijuana-strains/sativa-dominant-hybrid
/day-tripper

Devil's Lettuce
en.wiktionary.org/wiki/devil%27s_lettuce

Durban Poison
leafly.com/sativa/durban-poison

allbud.com/marijuana-strains/sativa/durban-poison

Dutch Treat
allbud.com/marijuana-strains/indica-dominant-hybrid
/dutch-treat

Gorilla Glue 4
allbud.com/marijuana-strains/hybrid/gorilla-glue-4

Jager
leafly.com/indica/jr

allbud.com/products/breeze-botanicals/flower
/37072/formerly-known-jaeger

Kandy Kush
leafly.com/hybrid/kandy-kush

Maui Wowie
leafly.com/sativa/maui-waui

allbud.com/marijuana-strains/sativa-dominant-hybrid
/maui-waui

Northern Lights
allbud.com/marijuana-strains/indica/northern-lights

leafly.com/indica/northern-lights

OG Kush
leafly.com/hybrid/og-kush

allbud.com/marijuana-strains/hybrid/og-kush

Permafrost
leafly.com/hybrid/permafrost

Predator Pink
leafly.com/hybrid/predator-pink

allbud.com/marijuana-strains/indica-dominant-hybrid
/predator-pink

Purple Chemdawg
leafly.com/indica/purple-chemdawg

Purple Hindu Kush
leafly.com/indica/purple-kush

allbud.com/marijuana-strains/indica/purple-kush

Snoop's Dream
allbud.com/marijuana-strains/hybrid/snoops-dream

leafly.com/hybrid/snoop-s-dream

Sour Diesel
leafly.com/sativa/sour-diesel

allbud.com/marijuana-strains/sativa-dominant-hybrid
/sour-diesel

Thin Mint Girl Scout Cookies
leafly.com/hybrid/thin-mint

allbud.com/marijuana-strains/hybrid/thin-mint-girl-
scout-cookies

SUPER CROPPING
royalqueenseeds.com/us/blog-bigger-cannabis-yields-
with-super-cropping-n815

VENTILATION: CIRCULATION
gardenandgreenhouse.net/articles/greenhouse-
articles/the-basics-of-greenhouse-ventilation

Magazines and Journals

If you're looking for information beyond online sources, these should give you a good start.

High Times: This magazine provides great information on how to grow marijuana, as well as the latest information on strains.

McPartland, John M., and Ethan B. Russo. "Cannabis and Cannabis Extracts: Greater Than the Sum of Their Parts?" *Journal of Cannabis Therapeutics,* vol. 1, no. 3/4, 2001, pp. 103–132.

Index

Numbers

A

B

C

T–U